€25 -

GW01605956

Western Frontiersmen Series
XXXIV

Horace Plunkett in America

An Irish Aristocrat on the Wyoming Range

by

LAWRENCE M. WOODS

THE ARTHUR H. CLARK COMPANY
An imprint of the University of Oklahoma Press
Norman, Oklahoma
2010

Library of Congress Cataloging-in-Publication Data

Woods, L. Milton (Lawrence Milton), 1932–
Horace Plunkett in America : an Irish aristocrat on the Wyoming range / Lawrence M. Woods.
p. cm. — (Western frontiersmen ; 34)
Includes bibliographical references and index.
ISBN 978-0-87062-394-3 (hardcover : alk. paper) 1. Plunkett, Horace Curzon, Sir, 1854–1932. 2. Irish—Wyoming—Biography. 3. Ranchers—Wyoming—Biography. 4. Wyoming—Rural conditions. 5. Aristocracy (Social class)—Ireland—Biography. 6. Diplomats—Ireland—Biography. 7. Political consultants—United States—Biography. 8. Ireland—Foreign relations—United States. 9. United States—Foreign relations—Ireland. 10. Ireland—Biography. I. Title. II. Series.
F761.P58W66 2010
978.7'02092—dc22
[B]

2009015253

Horace Plunkett in America: An Irish Aristocrat on the Wyoming Range
is Volume 34 in the Western Frontiersmen Series.

The paper in this book meets the guidelines for permanence and durability of the Committee on Production Guidelines for Book Longevity of the Council on Library Resources, Inc. ∞

Copyright © 2010 by the University of Oklahoma Press, Norman, Publishing Division of the University.
Manufactured in the U.S.A.

All rights reserved. No part of this publication may be reproduced, stored in a retrieval system, or transmitted, in any form or by any means, electronic, mechanical, photocopying, recording, or otherwise—except as permitted under Section 107 or 108 of the United States Copyright Act—without the prior written permission of the University of Oklahoma Press.

1 2 3 4 5 6 7 8 9 10

Contents

Illustrations

PHOTOGRAPHS BY GEOFFROY WILLIAM MILLAIS

Introduction

THIS IS NOT A full-length biography of Horace Plunkett. Margaret Digby wrote such a biography, following the outline Plunkett had prepared for the autobiography he never wrote, and, more recently, Trevor West wrote a carefully researched study of Plunkett, filling in many of the answers to questions raised by Digby. Yet my good friend Trevor was puzzled about some aspects of the western part of the Plunkett story, and there are sources in the United States that illuminate some of those aspects. Hence, this effort.[1]

In this work I focus on Horace Plunkett in the United States, and although the time he spent in that country was only a fraction of his life, his American experience left a lasting imprint on his later life. Indeed, there is compelling evidence that the Horace Plunkett who left the range in Wyoming was different from the Horace Plunkett who first arrived there a decade earlier. Horace was young when he went to Wyoming to take up his first real job away from home. He had a special incentive to succeed, for he expected that, as a younger son, his inheritance from his father would be modest. Without a doubt, the many troubles he encoun-

[1] As early as 1921, Houghton-Mifflin urged Plunkett to write his autobiography, and he apparently intended to do so, but the book never was written, at least in part because many of the sources he would have used were destroyed when his Irish home, Kilteragh, was ruined in 1923 by a bomb and subsequent fire (see chapter 14). Digby, *Horace Plunkett.* Also, Plunkett diaries, 16 Feb. 1921. Plunkett referred to Edward E. Lysaght as his biographer (without endorsement), but Lysaght's book, *Sir Horace Plunkett and His Place in the Irish Nation,* which appeared in 1916, is chiefly a discussion of Plunkett's involvement in the Irish Question.

tered in the American West had a maturing influence on him, and one can trace the progress of his thinking in the little summaries he often appended to his diary at the end of each year.

It is an open question whether the United States did more than hasten his maturation. Digby felt that the democratic environment softened Plunkett's aristocratic leanings, and perhaps it did, particularly when he dealt with the political structure of Ireland, but his class-conscious view of individuals continued to reflect his background and training in the Old Country. And certainly the roots of his great interest in agricultural cooperation sprang from the little store his father ran on the family estates, not from the self-reliant, independent western cattle industry.

Although the fruits of Plunkett's American investments were the only significant assets in his estate when he died, this money was, in large part, only a bare return of the capital he had invested, for a combination of economic and other factors conspired to limit the returns he received. When the *London Times* mistakenly supposed that Plunkett had died, its 1920 obituary, which took up an entire column in the paper, declared that he had accumulated "a considerable fortune" from ranching, a statement that must have brought an ironic smile to his face when he read it. In fact, none of his numerous investments in the United States yielded more than modest returns, and a number of them actually left him with losses.[2]

Nevertheless, Plunkett's American experiences enriched him in other ways, and he returned to the United States every year for so long as he was well enough to make the journey. At first, these journeys were to supervise his investments or to raise money for his Irish cooperative work, but later, he also tried to make his own contribution to the American scene. He saw the chance to use his understanding of agricultural cooperation to improve the quality of rural life in the United States, and he labored to improve public understanding between the British and American people. In this work, he came to know presidents and senators, as well as other Americans in all walks of life, who sought his opinions and often heeded them.

[2] Plunkett had refused an interview in New York at the end of 1919, and the reporter sent out word that he had died; this resulted in a profusion of obituaries, including one in the *London Times*, 1 Jan. 1920.

Whenever he became interested in a controversial topic, Horace Plunkett not only tried to understand both sides of the issue, but he also tried, often successfully, to open channels of communication with people on both sides. These relationships could have been used to dampen conflict but for the fact Plunkett also felt constrained to *criticize* both sides—the Plunkett version of even-handedness. Thus, when he criticized some of the policies of the Catholic Church in his first full-length book, he seemed at first to be puzzled at the firestorm from the clerics whom he had cultivated, not appreciating their deep resentment of such criticism—however well-intentioned—from a Protestant.

Plunkett's diaries cover the period from the beginning of 1881 until just before his death in 1932 and are the most significant source for this work, giving a remarkable insight into the man and the times he lived through. Yet they also give one the sense that he always supposed strangers might someday read them, and such diaries may not always be completely candid.[3]

Also, a reader of the diaries must be prepared for Plunkett's tendency to abruptly revise his opinions of people, for he might comment favorably on someone ("I liked the man") and then radically change his assessment, perhaps only a short time later. This may indicate a tendency to reach a snap judgment, but it also reveals a certain weakness in judging people. For whatever reason, it becomes obvious that Plunkett's record of choosing partners and trusted employees was a checkered one.

An important part of this story is to find the answer to the question of why Plunkett and his young friends left their comfortable homes in the British Isles to seek their fortune in the American West. Their assets for these ventures consisted of pounds sterling and unblinking confidence in their ability to learn a business they had no previous experience with. For most of these men, those assets were soon exhausted and they returned home, but Plunkett continued to engage in new businesses and to examine interesting American qualities and idiosyncrasies. Many of their reasons for

---

[3] On at least one occasion, Plunkett acknowledged his expectation that his diaries would be read by posterity. After recording a 13-hr trip in his new motor car, he wrote, "If ever this diary is read after I am dead & long buried, I suppose the performance of an amateur chauffeur getting a new car through a 257 mile run in the day will not appear very much of a feat. But I thought it creditable." Plunkett diaries, 20 May 1903.

coming to America, and for choosing their associates here, are found in the complex family relationships back home. To avoid tiresome recitations of these connections, I have placed them in footnotes so that the reader may learn something of the motivations at work.

Exchange Rates

Monetary transactions are quoted throughout in pounds sterling (£) and dollars ($), depending on the currency Plunkett was dealing with at the time. I have not converted these currencies for each transaction and instead offer this overview of the fluctuation of the value of the pound sterling in the period 1878–1932.

1878–1919	Gold standard years	$4.87
1920–1929		$3.66–$4.82
1931	Abandoned gold standard	$3.69

I

Horace Curzon Plunkett

The Dunsany branch of the Plunkett family considered themselves to be Irish at heart, but in fact, as Protestant noblemen with strong ties to the English rulers of the island, they could never be fully accepted within the Catholic Irish society. Even their well-meaning efforts to improve the lot of the poor were viewed with suspicion, and when Lord Dunsany, Horace's father, set up a cooperative store to aid the local farmers, he was accused of opening the store in order to buy his own food cheaper. Although Horace Plunkett was familiar with incidents of this sort, he never seemed to understand fully the implications they held for his own work. Indeed, R. A. Anderson, who worked for Horace for many years, said that Horace was English in everything except his heart, which was Irish.[1]

The Plunketts are an ancient Irish family, and some say they came from Denmark, whereas others prefer descent from William the Conqueror. There are several important branches of the family, including four headed by peers. The earldom of Fingall dated from 1649, the barony of Louth was created in 1541, and the barony of Plunket of Newtown was created in 1827. The Dunsany family branch is the one of interest in the current book, and Henry VI created the first Baron Dunsany in 1461. The twelfth baron Dunsany converted to the Protestant faith, and Lady Fingall said that he did so to save his horse, for the law did not permit any Catholic to own a horse worth more than £5. In any case, this religious anomaly among the Plunketts proved useful to the other branches of the

[1] Plunkett diaries, 14 and 21 April 1881; and Anderson, *With Horace Plunkett in Ireland*, 39.

family, for the Protestant Dunsanys could become "holders" of property for their Catholic relatives, and each year, Lord Dunsany had to renew his oath that he owned the properties of the Catholic Fingalls.[2]

Horace grew up in Dunsany Castle, which Hugh de Lacy built about 1190 on a site only three miles from the Hill of Tara, ancient seat of the High Kings of Ireland. The nearest neighbors were the Plunkett cousins of the Fingall branch. Horace Plunkett's father, Edward Plunkett, was an admiral in the British Navy who had served against pirates in the Mediterranean.[3]

At the time of Horace Plunkett's birth, Ireland had just passed through the horrors of the Potato Famine years of 1846–47. This natural disaster robbed the poor farmers of their ability to feed themselves, so that many died and many more fled the country, shrinking the population nearly two million from 1845 to 1851. Hunger was not the only blight on that unhappy, overwhelmingly Roman Catholic land, which was ruled sometimes badly, and at best indifferently, by the British Parliament at Westminster.

The Dunsany branch of the Plunkett family did not enjoy robust health, and Horace's brother Reginald died when Horace was ten, followed by his sister Julia two years later. It was then expected the barony of Dunsany would descend on Randal, but he also died, on Christmas Day 1883, leaving as the heir a brother John William, who had ruined an otherwise robust constitution with alcohol and emetics and now suffered from epileptic-like fits.[4]

Religion was a factor in the mental set of the family, for Horace said that his mother was a Low Church communicant in the Church of Ireland and "stoked the fires of Hell" in Lord Dunsany's mind, but it is impossible to briefly characterize Horace's religious inclinations. Ever the religious skeptic, although nominally a member of the Church of Ireland, he called its services "cold and valueless," and he also attended services of other denominations, albeit seldom with

[2] Hinkson, *Seventy Years Young*, 104. The Plunkett name is apparently derived from a corruption of the French *blanchet,* meaning white, and may refer to them as wearing white jennets or riding white horses.

[3] In a 1916 letter to Woodrow Wilson, Horace Plunkett said his father remembered rejoicing over the news of the victory at Waterloo. Horace Plunkett to Woodrow Wilson, 23 Feb. 1916, Link, ed., *Papers of Woodrow Wilson*, vol. 35.

[4] Plunkett diaries, 14 Feb. 1884. John William Plunkett died in 1899.

favorable comment. In Omaha, Nebraska, he attended a Unitarian service, which he described as a "wild lecture against inspiration of Bible. Style bad, facts good." His public work in Ireland brought him in frequent contact with the Roman Catholic clergy, but he was critical of the Roman Catholic mass, calling it "muttered and mumbled," poorer even than the Church of Ireland service.[5]

Lord Dunsany was aloof, and his distant cousin Lord Fingall once said, "You would never think of putting your hand on his arm." Horace developed similar traits, and R. A. Anderson, a long-time employee of Horace's, declared that he never addressed his employer by his Christian name, although Horace did instruct him to drop the "Mr." Horace had a stoic quality, and Digby regarded it as the "real basis" of his life. When Horace was twenty-seven, he remarked that he was not happier than when he was a child but "certainly less apprehensive of future sorrow," and in one of his last speeches he said, "The virtue lies in the struggle, not the prize."[6]

There are only a few occasions when Horace recorded moments of happiness in his diaries, and when he did, it was as though he had to apologize for good feelings. Thus, at the time of his winter ball at Dunsany Castle in the winter of 1881, he said that he had been "foolishly—imprudently—happy." A week later, when he had a bad nightmare involving murder and sudden death, he solemnly attributed the dream to the "natural effects" of the joy he had known during the week.[7]

On the one hand, Horace Plunkett's background indelibly marked him with an innate aristocratic snobbery, but he also harbored many egalitarian principles. Thus, he was critical of the selfishness of some members of his social class, saying they had no real ambition and wanted only to gratify their pleasures. These contradictory influences turned up repeatedly, and although his democratic tendencies were undoubtedly bolstered by his American experiences, the older shadow of his upbringing never completely left him.

When Horace was born on October 24, 1854, he received the middle name of Curzon, a family connection of his mother, who

---

[5] Ibid., 15 May 1881 and 26 Nov. 1881.

[6] Digby, *Horace Plunkett*, 302–303; and Plunkett diaries, 17 April 1881.

[7] Plunkett diaries, 12 and 19 Feb. 1881 and 24 Oct. 1884.

was the daughter of Lord Sherborne. Lady Dunsany told her husband that Horace was a "young ruffian," whose strong will distinguished him from her other children. It was just as well for him to be self-reliant, for as the fourth son in his family, Horace could not expect to inherit his father's title.[8]

There are few known details regarding Horace's education. In 1868 he attended Eton, that great private school (called "public" in the British usage), where he enjoyed the privilege of being in Oscar Browning's house. Browning never married, and his house was presided over by his mother and sister. Forty boys aged fifteen to nineteen years were fed luxuriously in an atmosphere that made rough-housing unthinkable and was intended to emulate conditions in the boys' own homes.

Discipline by the kind-hearted Browning was much more humane than in other Eton houses, and the presence of a large staff of servants limited—but did not eliminate—the system of fagging, where the younger boys ("fags") served the older students. A position in Browning's house was in such demand that parents sometimes entered their boys' names with the school at birth. Although Horace often mentioned schoolmates at Eton, we do not know how he felt when he was there, but a cryptic 1884 diary entry, in which he said he was weak and miserable during a bitter winter sixteen years earlier, may refer to that experience.[9]

Horace left Eton in 1872 to study with a tutor at Old North, at Brighton, and in 1874, he entered University College, Oxford. While at Oxford, he played chess against John Neville Keynes and earned the reputation as a chess player of the first rank. He continued to be a strong player until nearly the end of his life, often even playing blindfolded.[10]

---

[8] Digby, *Horace Plunkett*, 5, 13.

[9] Browning was dismissed from Eton in 1875, in part because the headmaster thought Browning's relationship to George Nathaniel Curzon (the future Viceroy of India) was more intimate than it ought to have been. Curzon came to Eton in 1874 at the age of 15. Wortham, *Victorian Eton and Cambridge*, 62–65. Plunkett's fag at Eton was John Charles Kennedy, who was born 23 March 1856, and came to the school in 1871. In March 1881, Horace went to see Kennedy, who had married in 1879 and had succeeded to his father's baronetcy in 1880. Plunkett said that Sir John was "glad to see me and we had lots of talk about old times." Plunkett diaries, 3 March 1881 and 7 Feb. 1884.

[10] In the spring of 1925, Plunkett returned to Eton and saw his name on the list of those who left in 1872; he also saw his room and found it little changed. Plunkett diaries, 17 May 1925.

Digby, Horace's biographer, said that as a young man, Horace worked hard and played hard, going to bed at 6:00 A.M. after dancing all night and rising again at 7:00 A.M. Daisy Fingall, who was married to Horace's distant cousin Lord Fingall, said that her first impression of Horace was of someone "very pleasant looking and gay," but she also remarked that he seldom smiled in the latter days of his life. R. A. Anderson remarked that praise from Horace was infrequent. Indeed, Horace had few intimate friends, either male or female.[11]

He never married, and in his early diaries there is little mention of the possibility of marriage, except in the negative sense, when he was concerned that the mental anguish he often suffered might be passed on to his children. Certainly, he enjoyed the company of attractive young women and accepted with alacrity when invited to dances, although Daisy Fingall remembered that when she first met Horace, at a ball at Killeen Castle, he danced as badly as Lord Fingall.[12]

Daisy Fingall was a special person in Plunkett's life in Ireland. In the spring of 1883, she married Lord Fingall, whose somnolence stemmed from disposition, not age (he was actually five years younger than Horace). Fingall was no match for the lively Daisy, and she gladly took up the role as Horace Plunkett's social hostess. Surely this was a welcome outlet for her, but there is no proof in available records as to whether there was a more intimate relationship between Horace and Daisy.[13]

Although Horace's handshake was limp, even flabby, he had a strong personality and tended to dominate most of his closer relationships with other people. Alexis Roche, Plunkett's early business partner in Wyoming, said that Horace was strange—"like nothing on earth"—and remarked on his obsession with detail, saying, "Don't let Horace near anything you have written. He'd amend an order for a bag of coal." This fascination with detail also drew Horace to the science of accounting. He taught himself double-entry bookkeeping and soon was proficient at it.[14]

---

[11] Hinkson, *Seventy Years Young*, 87

[12] Ibid.

[13] Plunkett's first impression of Daisy Fingall was distinctly unfavorable, but this view soon changed. Plunkett diaries, 21 June 1884.

[14] Anderson, *With Horace Plunkett in Ireland*, 1, 154.

Horace was not an impressive speaker, and his Irish audiences complained that he had trouble pronouncing the letter "r," and he spoke "Englified." (Of course, for them he also suffered from being the son of a landlord, and a Protestant.) Moreover, he always had trouble remembering faces, so that he often walked by people who had been introduced to him before—perhaps many times. A fellow member of the Kildare Street Club, on being asked whether he knew Horace Plunkett, replied that they had been introduced "at least eleven times." This difficulty, coupled with his naturally aloof bearing, often gave Horace a reputation as more of a snob than he really was.[15]

Plunkett's health troubled him throughout his life, and Digby declared that Plunkett came to the United States because he was afflicted by tuberculosis, "the family malady." The doctor told him both his lungs were affected and said that his only chance to recover would be to go to South Africa or Colorado. Horace chose Colorado, although he stopped there only long enough to buy a pair of mules and a wagon before going north to Wyoming Territory.[16]

Although he never showed signs of tuberculosis in the American West, that region did not work a miracle cure for Plunkett's health either. Even though he was 5'10" tall, he weighed only 130 pounds or less when he was out on the range in the West, and when he reached 136 pounds, it was an occasion to remember. Despite his rather fragile physique, he did not pamper himself, walking about rapidly, talking all the while, in general pushing his body beyond its limits, and bringing on bouts with diarrhea. A persistent ringing in his ears bothered him at intervals all his life, and he was convinced that this could be stopped only by a faultless diet, which he despaired of finding in the American West.[17]

Medicine was a rough and tumble business in those days, and doctors tried remedies that would not be considered appropriate today, as in the fall of 1883, when a doctor in Cheyenne gave Plunkett opium for diarrhea. Later, other doctors abused the wonderful possibilities of X-rays, and a nasty burn in 1916 permanently

[15] Ibid., 3, 7, 39, 153–54.

[16] Robinson, *Lady Gregory's Journals*, 236.

[17] Plunkett's health did not improve after he left the American West, and at the end of 1891, he weighed only 123 pounds, "naked." His chronic diarrhea and other physical problems plagued him for the rest of his life. Plunkett diaries, 4 Dec. 1891.

undermined his already weakened health, setting him on the steady decline that ended with his death.

Poor health also created mental problems for him, and when he went to the ranch in the spring of 1881, he arrived so unsettled by the journey that he had to spend several days at the ranch house before he was able to go out on the range to work. He received conflicting advice from doctors, as one suggested a diet with many vegetables, whereas another told him he needed to eat more meat. A doctor in Douglas, Wyoming, treated Plunkett for a sore throat with a combined swallow of bromidium, iron chlorate of potash, and glycerin. He grimly noted in his diary, "Suppose it's poisonous."[18]

Plunkett's decision to go to the American West was undoubtedly influenced by the doctor's advice, but his reason for leaving Ireland was to make his fortune so that he could afford to live as he had been accustomed to in his father's castle. This was the typical dilemma for the younger sons of titled aristocrats. Barred from inheriting the bulk of their fathers' estates because of the system of entail on the eldest son, they were supposed to find "acceptable" employment in the church, the military, or the government. Ranching in the American West now became an additional form of acceptable work, if it could be conducted on a suitably grand scale.

Horace was such a younger son. When Horace left for America, his brothers Randal and John were still alive as was John's son, who had been born the year before; hence, the probability that Horace would succeed to the barony and its assets was extremely remote. Although Lord Dunsany later favored Horace disproportionately in his will, Horace never learned of that good fortune until his father's death. His work and investments in America were always undertaken to provide financial support for some of his cooperative stores as well as to give him a dependable financial future.

[18] A diary entry written on the ranch gives Plunkett's "naked weight" at 9 stones, 4 pounds (130 pounds), and his height at 5'10". Plunkett diaries 16, 23, and 26 June 1884; 17 Aug. 1884; and 5 June 1887.

2

Ranching in the American West

As noted in chapter 1, Plunkett did go to Colorado in 1879, as the doctor recommended, but he stayed there only long enough to buy mules and a wagon for the trip north to Wyoming Territory, where he intended to engage in cattle ranching. Wyoming Territory was created in 1868, mostly from the western end of Dakota Territory (but including some pieces of Idaho and Utah territories as well), and a large part of the territory was still a remote frontier in 1879. The Powder River Basin in the northeastern corner of the territory, east of the Big Horn Mountains, was particularly remote, for it had been the unceded Sioux hunting grounds until after the Custer debacle on the Little Big Horn in 1876, only three years before Plunkett and his friends arrived.

Communication with the Powder River Basin was by horse or wagon, and everything shipped from the civilized world had to be sent to the Rock Creek station on the Union Pacific and then taken north some 200 miles over roads often impassable in all seasons. When the young British aristocrats arrived in the Powder River Basin, there was no local government in the northeast, for neither of the two counties created there by the 1875 territorial legislature were organized before 1881. Buffalo still roamed on the hills of the Powder River Basin, and as late as November 1881, Plunkett saw a large herd there.[1]

---

[1] Plunkett diaries, 1 Nov. 1881. In 1877, the territorial legislature wanted to rid themselves of an unpopular judge, whom they could not legally dismiss, and hit upon the plan to assign him to the two unorganized northern counties, which had few non-Indian residents. Judge William Ware Peck did not go to his new district, but the word "sage- (*continued, next page*)

The range cattle industry owed its beginnings to the discovery that cattle could survive on the plains in the winter without supplemental feed. Travelers on the great emigration routes across Wyoming abandoned exhausted or "spent" livestock that could no longer keep up with the wagon trains, leaving the animals to fend for themselves. To the surprise of all (including the hapless animals), in the spring many of these animals had survived and even grown fat by eating the dry, "cured" grass. This unexpected aspect of the High Plains identified a cheap food source of free-ranging animals on the public domain. A market for the meat appeared when the construction crews came to build the transcontinental railroad, and the range cattle industry was born.

Completion of the railroad also opened the High Plains to larger markets in the eastern cities, and fortunes were made quickly in cattle ranching. Moreton Frewen established British ranching in the Powder River Basin following a trip he made to the Texas Panhandle in the spring of 1878, where he saw and was impressed by the JA Ranch of Goodnight and Adair. In the fall of that year, Moreton and his brother Dick came to Wyoming on a hunting trip, after which they rode over to the Powder River Basin to locate their new cattle ranch, where they would graze cattle on free public domain grass.[2]

Information from the Frewen brothers drew other young aristocrats to the area. James Boothby Burke Roche and his younger brother Alexis, both younger sons of Lord Fermoy, sailed from Liverpool on the *Abyssinia* and landed in New York (accompanied by their servant) on May 7, 1879. They made their way across the country to Cheyenne in a leisurely fashion, checking in at the Railroad Hotel on May 26. From there, they went north to the Powder River Basin to settle near the Frewens.[3]

On July 11, 1879, Plunkett landed in New York on the *City of Chester*, accompanied by Edward Shuckburgh Rouse-Boughton (the second son of a baronet) and Augustus "Gussie" Villiers

(*continued from previous page*) brushing" entered the Western vocabulary, to describe the exiling of a person. The animal Americans referred to as the "buffalo" was the American bison, a relative of the European bison, and not a true buffalo.

[2] The JA Ranch, located in the Texas Panhandle, was owned by Charles Goodnight and John George Adair from Ireland. There were a total of six men in the Frewen hunting party, including James Boothby Burke Roche and Gilbert Henry Chandos Leigh, who later lost his life in the Big Horn Mountains.

[3] *Cheyenne Daily Leader*, 27 May 1879.

Briscoe. They took the train to Denver, where Horace bought mules for the trip north to join the Roche brothers, who had arrived on the range in May. Plunkett, Boughton, and the Roches established their ranch in Wyoming's Powder River Basin, near the Frewen brothers' ranch.[4]

The Frewens and the Plunkett group were only the first of a flood of British investors in Wyoming. The impetus for the new British investment in the area came from the publication in 1879 of a comprehensive study for the British government, which concluded that annual returns of more than 33 percent were possible on the public domain in the United States. Soon other British investors arrived to enter this strange business that required almost no capital investment.[5]

These young British gentlemen pursued an enterprise fundamentally different from that of the large foreign ranches already established in Texas. Texas cattle ranchers early became substantial landowners and fenced their ranges to keep other herds out. The rise in land values that followed the settlement of the Texas cattle ranges eventually paid substantial dividends to patient shareholders in those companies, but the initial investment was heavy. Ranching on the free range of the public domain was entirely different.

The public lands on the High Plains did not receive enough rainfall to support crops without irrigation, which was an expensive undertaking for settlers. A homestead entry could not be filed before the land was surveyed, and when the early ranches were established on the northern plains, most of the land was still unsurveyed, making it impossible for anyone to file on a homestead. Because no one owned the grass, it was available without cost to anyone turning cattle out to graze on it. Although free grass was a great bounty for a time, the risk of overcrowding was also great, because anyone could bring a herd to graze on the same range. Nevertheless, during the years when grass was plentiful, ranchers only needed money to buy cattle and to hire cowhands, and profits were exceptional.

The 1882 boom in Wyoming ranch sales, to foreigners and others,

---

[4] Briscoe was twenty-two and Boughton twenty-one. The Rouse family came to England with William the Conqueror, and their baronetcy dated from 1641. Robinson, *Lady Gregory's Journals*, 236.

[5] The 1879 Royal Commission was headed by the Duke of Richmond, Lennox, and Gordon, and the report was delivered in 1881.

infused £6 million of Scottish and English money into the territorial economy and caused prices to jump 25 percent. The financial base of the industry seemed so sound that cattlemen boasted that livestock were a greater and steadier source of wealth than all the gold and silver mined in the mountain region.[6]

The Frewen brothers and the three newcomers to the Powder River ranged in age from Boughton as the youngest at twenty-one to Richard Frewen at twenty-seven, and all five were unmarried. The Frewen family did not have a title, but the others were younger sons of peers and a baronet. Their formal education may have given them some understanding of Latin and Greek, but they had no real experience running a business. However, they were all supremely confident that their background and intelligence would easily make them a match for the unsophisticated and often unlettered folk they met in the American West. That assumption would prove to be a costly error.[7]

Soon, Alexis Roche, Horace Plunkett, and E. S. R. Boughton formed the ranching partnership of Plunkett & Roche. Moreton Frewen claimed that he had arranged the ranching partnership, and this may well have been the case, for Moreton, who could certainly claim first arrival status, was then playing the role of the elder statesman of Powder River ranchers. Because there are no Plunkett diaries for this period, there are no details available, but formal operations under the partnership began in the middle of October 1879, as Plunkett's accounts dated from that point.[8]

Horace Plunkett was shorter and slighter than Alexis Roche, who was a large man, but Plunkett's personality dominated most of his personal relationships, both in the partnership and with

[6] *Cheyenne Daily Leader,* 3 April 1883.

[7] James Roche and Briscoe did not invest in ranching.

[8] Alexis served as a gentleman at large in the household of the Duke of Marlborough, after the duke was appointed lord lieutenant of Ireland in 1876. *Cheyenne Daily Leader,* 27 May 1879. James and Alexis's brother Edward died on 1 Sept. 1920, and James (the future great grandfather of Princess Diana) succeeded to the barony for two months until he also died, on 30 Oct. 1920. Also see Moreton Frewen to Clara Jerome, 28 Nov. 1880, Frewen collection, American Heritage Center, University of Wyoming (hereafter cited as Frewen collection). In a letter to the *Cheyenne Daily Leader,* dated 19 Sept. 1879, the unidentified writer (almost certainly Moreton Frewen), announced that Horace Plunkett, son of the "celebrated" admiral, was building a ranch at the head of the Powder River in partnership with Alexis Roche. *Cheyenne Daily Leader,* 23 Sept. 1879; and Plunkett diaries, 26 Aug. 1881.

many other people. No important decisions in the ranching partnership were made without his consent, and he maintained a similar position in nearly every venture that he undertook for the rest of his life. Although the partnership was initially styled "Roche & Plunkett," Horace shortly began referring to it as "Plunkett & Roche," and this latter style remained in use, both in his diaries and publicly, until the partnership was later folded into the Frontier Cattle Company.

Alexis was afflicted by a bad stammer, but Horace admitted that he had a certain flair as well as a memory for anecdote and a ready, "if rude," wit, which made Alexis popular among Americans in the West. Nevertheless, the partnership was barely a year old when Plunkett began criticizing the character defects of his partners. He blamed Alexis's problems on the fact Alexis had not attended a public school, where he would have been "fagged" and kicked. Although Horace was at first tolerant of these perceived shortcomings, later he lost patience with Alexis. Boughton was an even more difficult case, and Horace attributed Boughton's problems to the fact that he had grown up surrounded by old people and assumed that, at twenty-three, he possessed an equivalent wisdom.[9]

Other ranches appeared near those of the Frewens and Plunkett & Roche. On the middle fork of the Powder River, Thomas Willing "Willy" Peters and his partner, Walter C. Alston, established the Bar C Ranch, and they were joined by two British aristocrats as "junior" partners. Ralph Stuart-Wortley was the nephew of Lord Wharncliffe, and Geoffroy Millais was the younger son of the noted painter Sir John Everett Millais, who was soon to be created a baronet. Peters soon acquired the nickname "Twice Wintered," because he did not leave the ranch in the fall as the other owners did. In the fall of 1883, Stuart-Wortley, then nineteen, and Millais, twenty, fancied the idea of spending the winter on the ranch, but Peters only permitted Stuart-Wortley to stay, sending Millais home.[10]

---

[9] Plunkett diaries, 1 Sept. 1881.

[10] After spending a day with young Millais, Plunkett called him a "dreadfully dull boy." Moreton Frewen to Clara Frewen, 12 Aug. 1883, Frewen collection; and Plunkett diaries, 27 Aug. 1882 and 18 Sept. 1883. Ralph Granville Montagu Stuart-Wortley was born 4 July 1864, and Geoffroy William Millais was born 18 Sept. 1863. Geoffroy's sister Alice was married to a cousin of Lord Wharncliffe. Gladstone created John Everett Millais a baronet in 1885, the first artist to be so honored.

It was a formidable task for the British aristocrats to establish ranches in the Powder River Basin. As noted previously, all the trappings of civilization had to be brought by Union Pacific rail to the Rock Creek station west of Cheyenne and then taken north by stage or wagon. The British gentlemen usually stopped in Cheyenne on the way to or from their ranches, and in 1880 the Cheyenne Club was established there to introduce a bit of luxury and privilege to the West. Impetus to organize an exclusive club came from the local ranchers, although Plunkett and both Frewens were among the early members. Club membership was limited to 200, and drunkenness and profanity were grounds for expulsion as was "any act so dishonorable in social life as to unfit the guilty person for the society of gentlemen."[11]

The journey to reach the ranch on the Powder River from Rock Creek was a jarring 200 miles, requiring thirty-seven hours on the stage when the roads were passable, which was by no means always the case, even in summer. An August stage journey was a mixture of heat, dust, cramps, aches, alkali water, and filthy food, mostly canned, and if the passenger tried to sleep, he was likely to be awakened when he was thrown up in the air, perhaps hitting his head on the roof of the coach.[12]

Although Moreton Frewen displayed a certain frontier elegance in his big house (afterward nicknamed "Frewen's Castle"), the Plunkett & Roche ranch house accommodations were not luxurious. When Plunkett returned to Ireland for the 1880–81 winter, he was thankful to be back in a "comfortable" house, as compared with the ranch, where the only furniture still consisted of three-legged "semi-bottomed" chairs.[13]

Aside from this minimal investment in housing, ranch owners had to buy only cattle and a string of horses. A reliable foreman was needed, as he was the one who hired (and fired) the cowboys and in general directed the work on the ranch. Choosing a foreman was serious business, and the British owners usually sought the advice of their countrymen on the range before hiring one.

---

[11] Richard Frewen was an incorporator of the Cheyenne Club, but all the other charter members were local ranchers. Minutes of the Cheyenne Club, 5 and 22 June 1880 and 16 and 21 Sept. 1882, American Heritage Center, University of Wyoming.

[12] Plunkett diaries, 7 Sept. 1881.

[13] Ibid., 17 Jan. 1881, 30 May 1881, and 6 Oct. 1886.

Numbers of cowboys could be found in the western saloons, but the quality was most unpredictable.

Most of cowboys were hired only for the twice-yearly roundup work, after which they were laid off, and the winter contingent consisted only of the foreman and a few line men. Details on the early staffing of Plunkett & Roche are not available, but there is some information on Plunkett's later ranching partnership with Windsor & Coble. This organization employed a dozen or more men, plus hay hands, during the roundup seasons but only two or three in the slack period. Before 1885 (when wages were cut), cowhands were paid $40 per month, and more experienced men could make $45 or even $50, whereas the foreman could make $900–1,200 per year.

The men were paid twice a year, after each roundup was completed. Those cowhands who were lucky enough to be kept on the payroll between roundups could charge the cost of their tobacco or a new slicker to their account with the ranch, and for their cash needs, they borrowed from each other or from the foreman. When settlement time came, the charge accounts and the borrowings were deducted from the total earnings, and if the man was lucky and not too extravagant, he would receive a net payment.[14]

In the years before 1885, cowboys who were laid off between roundups or for the winter could "ride the grubline"—visiting the big ranches for a few weeks in turn, where they could sleep in the bunkhouse and enjoy free meals. This accommodation for unemployed cowhands was merely an extension of the early western custom of hospitality to all travelers on the range. For the unemployed cowhands, this policy operated as a sort of private welfare system and, not incidentally, made it more certain that a pool of cowhands would be ready for hire in the next roundup. Later, in the lean years, this spirit of benevolence evaporated, and Horace Plunkett was involved in those later policy changes.

To finance their ranching investments, the Frewen brothers brought to the West the remnants of their inheritance from their father,[15] and Alexis Roche and Boughton similarly invested funds

[14] Ranch managers, such as Fred G. S. Hesse, earned far more than the foremen.

[15] Moreton Frewen, who was born in England in 1853, joyfully indulged his love of fast horses and beautiful women before he came to Wyoming in 1879; his older brother Richard was more conservative.

their families provided. Plunkett also got by with advances from Lord Dunsany—enough money to start the business but not enough to carry him through a truly disastrous year or several years of lesser misfortune. As a consequence, Horace continually had to return to his generous father to receive further outright grants or "loans" (the latter were forgiven when the old man died).

Horace Plunkett was only 25 when he came to the American West in 1879, and he could not have been prepared for the demands sometimes placed upon him by life on the range in that frontier region. Although he had hunted on many horses in Ireland, Plunkett did not at once try to master the art of riding a bucking bronco. In the spring of 1885, he decided to try, and he made his usual analysis of the results. He found he could take the bucks without being dislodged from his seat, but the rearing of the horse was more than he could stand, and he "got a shake on the yard."[16]

Immediately after the first British ranches were established in northeastern Wyoming, friends and relatives from the mother country came there in some numbers, sometimes by invitation and sometimes totally unbidden. Although transportation in the range country was a rugged undertaking, these visitors could easily reach Cheyenne. The transcontinental railroad attracted travelers from all over the world, who used this route to avoid tiresome and dangerous sea voyages, and those who crossed the continent on the train often stopped in Cheyenne.[17]

---

[16] Plunkett diaries, 14 May 1885.

[17] In the days before the Canadian Pacific was built across Canada, Cheyenne was periodically treated to the pageantry of the entourage of the governor general of Canada, who could not otherwise easily reach the western part of his country. Plunkett even met King Kalakaua of the Sandwich Islands (1874–91) on the train during one of his journeys to the East.

3

Frenetic Wyoming Rancher, 1881

NOT MUCH IS KNOWN about Plunkett's first year in ranching, because his diaries did not commence until 1881. The year 1881 was a busy one for Plunkett—not because he devoted himself single-mindedly to learning the business of ranching, but because he spent a good deal of time investigating an amazing number of other ideas that came to his mind. In the ranching business, a deal he contracted for in 1880 claimed much of his attention in 1881 and beyond.

In July 1880, Plunkett negotiated a contract to buy cattle from the Searight Brothers' ranch in Wyoming Territory. The contract was for a major addition to the Plunkett & Roche herds—more than 3,000 head of cattle at a total cost of nearly $54,000. Subsequent difficulties with this contract not only exposed the first major friction among the Plunkett & Roche partners, but also gave Horace a hard lesson in the problems British gentlemen encountered when dealing with Wyoming ranchers in Wyoming courts.

The contract was between two partnerships, Roche & Plunkett, said to consist of Alexis Roche and Horace Plunkett (ignoring the Boughton interest), and Searight Brothers, consisting of Gilbert, Francis, and George. However, the negotiations took place between Horace Plunkett and Gilbert Searight, the only Searight brother then living in Wyoming.[1]

Gilbert Searight was born in 1834, making him a generation older than Horace Plunkett, who was not yet twenty-seven. The

[1] Searight Brothers had their cattle on Poison Creek, not far from the Plunkett & Roche operation on the Powder River.

differences between the two men did not end with their ages, for the Searights of Carlisle, Pennsylvania, had little in common with the younger son of an Irish baron. Although the range cattle business was young in the Wyoming Territory, Searight had been in the cattle business in Texas in 1870 before coming to Wyoming. For his part, Horace Plunkett was trying hard to learn the business.

The contract called for delivery of both breeding stock and animals for future sale, and it was intended as a major element in the launch of the new cattle business for Plunkett & Roche. Unfortunately for Plunkett, the contract did not explicitly recite this objective. The contract called for payment of $40,000 on or before the delivery of the cattle and the balance six months later. Plunkett & Roche actually paid $1,000 at the time of signing the contract and a further $20,000 on August 30, 1880.[2]

The agreement was very detailed, calling for delivery of specific numbers of cattle in various classes. Unfortunately for both parties, the cattle described in the contract were a part of herds somewhere out on the range, and no one knew in detail how many of each class were out there. Nor did anyone know how many of the animals out on the range would be found and gathered by the roundup crews. Nevertheless, Plunkett expected that if Searight did not deliver precisely what the contract called for, he, Plunkett, would be able to enforce his rights in the courts. This hope proved to be vain.[3]

Unfortunately, when the Searight Brothers sent their roundup crews out to collect the cattle for delivery, they did not find enough cattle to fill the contract. Precisely how or when the Searights told Plunkett & Roche about this situation is not known, because the details later advanced by the two sides in the subsequent lawsuit

---

[2] The contract, dated 24 July 1880, called for four classes of animals. The breeding stock consisted of 666 cows, 1,000 two-year-old heifers, and about 350 yearling heifers, and the animals for future sale consisted of 750 two-year-old steers for sale the following year and about 350 yearling steers. The contract is found in the files of the Laramie County District Court, with the petition of Roche & Plunkett v. Searight Brothers, Case No. 435, filed 20 Dec. 1880.

[3] Gilbert Alexander Searight was born in Carlisle, Pa., on 5 Oct. 1834, George Peter Searight was born in South Middleton, Pa., on 28 May 1838, and Francis William Searight was born in Carlisle, Pa., on 1 March 1827. Gilbert was living in Burnet, Tex., at the time of the 1870 census, where he was engaged in the cattle business. The other two brothers continued to live in Pennsylvania.

are sharply at variance. Nevertheless, it is clear that something was said about this on or about November 1, because the Searights claimed they entered into a new contract with Plunkett & Roche on that date—an allegation that Plunkett denied. In any case, on November 3, the Searights delivered 781 animals called for in the July contract, leaving a shortage of 2,335 head, but they also delivered 109 three-year-old steers and eleven horses not called for by the original contract.[4]

As soon after the Searights failed to deliver the cattle required by the contract, Plunkett decided to sue them in the Cheyenne district court, and to do so he needed to hire a local lawyer. The firm he selected was Brown & Potter, consisting of Melville C. Brown and Charles N. Potter. Because Brown lived in Laramie, Plunkett dealt with Potter, who later had a distinguished judicial career in Wyoming. Potter, then twenty-eight, was born in New York and graduated from the University of Michigan law school in 1873. He practiced law in Michigan until 1876, when he came to Cheyenne, where he was appointed city attorney at the age of twenty-five. In 1880 Potter was elected county attorney for Laramie County, and this record of achievement undoubtedly caused Plunkett to hire him for the Searight case. In mid-November 1880, Plunkett gave Potter the relevant information to file the lawsuit and then returned to Ireland for the winter, as was customary with all of the British cattle ranchers.[5]

Potter drafted the petition against the Searights, asking for damages of $25,000, including lost profits of $6 per head on the 2,335 head that had not been delivered by the Searights. The petition claimed that the Searight Brothers knew Roche & Plunkett needed the cattle to stock their range, and that there was no market

---

[4] The deliveries of 3 Nov. comprised 361 head of yearling cattle, 150 two-year-old steers, 109 three-year-old steers, 114 two-year-old heifers, and 156 cows; these animals were accepted by Plunkett & Roche. On the same day, the Searights also delivered eleven horses.

[5] *Cheyenne Daily Leader*, 14 Nov. 1880. Charles N. Potter was born in Cooperstown, N.Y. on 31 Oct. 1852, and practiced law in Grand Rapids, Mich., before he came to Cheyenne. He was later a member of the 1889 Wyoming constitutional convention, was elected attorney general in 1891, and was elected to the Wyoming Supreme Court in 1894, becoming chief justice in 1897. Horace Plunkett left Cheyenne on 14 Nov., en route to Ireland.

where suitable replacements could have been purchased after the default in deliveries.[6]

The Searights hired William Wellington Corlett to defend them against Plunkett & Roche, and he was certainly no lightweight either. Corlett, then 38, had been Wyoming territorial delegate to Congress from 1877 to 1879, and he turned down appointment as chief justice of the Wyoming Supreme Court to return to the practice of law. In the first of a flurry of motions in the Searight case, Corlett attacked the cornerstone of the Plunkett & Roche lawsuit, which was the allegation that the Searights had deprived Plunkett & Roche of lost profits of $6 per head on the 2,000 head that had not been delivered. Corlett contended this allegation was irrelevant. The parties then waited to see how the judge would rule.

In Ireland at the beginning of 1881, Plunkett caught up on the lifestyle he had to forego while in Wyoming, riding to the hunt sometimes five times a week and preparing for his sister's wedding and for his annual ball at Dunsany castle. Despite the social whirl, Plunkett still had time to think of American business. He talked with Walter Bulwer, who wanted to send his son to work on the ranch, and at the end of the month, Horace agreed that young Bulwer could join him in Wyoming. Next, he convened a business meeting with Boughton and Alexis Roche, who had also gone home for the winter. Plunkett blamed Alexis for the Searight troubles, calling it a "fearful mess," without stating why. (Because Plunkett had signed the contract, he may have blamed Alexis for accepting delivery of the motley herd of animals from the Searights.) In any case, the really important agenda item for the partners was the need to raise more money, which they all agreed to do.[7]

Weather reports Plunkett received from Wyoming told of the

---

[6] It is impossible to reconcile the allegations in the petition with the monetary amounts claimed. Plunkett & Roche claimed to have paid a total of $21,000 to the Searights and to have received cattle specified in the contract to a total of $12,589, plus other animals not called for worth $2,728. Leaving aside the last item, they appear to have overpaid under the contract by $8,411, which when added to the $14,010 claim of $6 per head on the deficiency, would give a total of $22,421, not the $25,000 requested. In any case, Plunkett & Roche accepted the horses and three-year old steers worth $2,728, which would have to be paid for.

[7] Plunkett diaries, 11, 22, 26, and 31 Jan. 1881; 16 and 26 Feb. 1881; 19 March 1881; and 7 April 1881. Captain Raymond Parr married Constance Lavinia Harriet Plunkett on 7 April 1881.

hard winter of 1881, in which large numbers of cattle died in Colorado and Wyoming. Unfortunately, the local newspapers in Wyoming only carried stories about conditions in Nebraska and Montana, but little about the Wyoming ranges. The exception to this statement was a pair of stories written by Plunkett's friends in the Powder River country that appeared in the *Cheyenne Leader* on the same day in January. One, signed "Pioneer" and undoubtedly written by Moreton Frewen, told of snow drifts six to eight feet deep and of 25 percent to 50 percent cattle losses. The second, written by Thomas Willing "Willy" Peters, a neighboring rancher, said the weather was moderate, with no snow and plenty of feed for the cattle.

The Frewen letter was undoubtedly written to discourage other ranchers from crowding the northern ranges, but Peters's report must also have been more than a little optimistic. We know the winter of 1880–81 was severe, but the 1881 cattle losses were less severe than in the bad winter of 1886–87, because the range in Wyoming was not as crowded as it became in the later period. For Plunkett & Roche, the irony of that winter was that the 2,000 head of cattle the Searights failed to deliver were not at risk to winter storms.[8]

It was May 7 when Plunkett again landed in New York, but he was not ready to go directly to the ranch. First, he wanted to get some legal advice on the Searight case from a New York lawyer. The man he chose was Charles Winthrop Gould, whose business was in Wall Street. Gould could not have known much about the cattle business, although he had worked at sheep raising before he went to Columbia law school. Rather than call on Gould at the office, Plunkett followed what he thought to be New York etiquette ("I am proficient at it now") by first meeting Gould and his new wife socially. The next morning he took the Searight papers to the lawyer's office, where he got "much advice" and was told the case was "impregnable." This favorable advice soon proved wildly optimistic.[9]

After leaving New York, Plunkett stopped in Omaha, Nebraska, where he was introduced to the Right Reverend James B. O'Connor,

---

[8] The letter signed "Pioneer" was postmarked 17 Jan. from Fort McKinney. Willy Peters was the partner of Walter C. Alston. *Cheyenne Daily Leader,* 23 Jan. 1881. Also, Plunkett diaries, 9, 10, and 15 March 1881.

[9] Plunkett diaries, 7–9 May 1881. Gould was later a cofounder of the firm of Gould and Wilkie.

Catholic bishop of Omaha. The two men talked about land speculation in Omaha, not religion, and this meeting marked the beginning of a long relationship between the two men. At the end of the month, Bishop O'Connor wrote to Plunkett in Wyoming, recommending the purchase of Omaha lots. Plunkett did have a religious experience in Omaha—although not a Catholic one—when he attended the Unitarian service mentioned earlier. Also in Omaha, a reporter from the *Bee* asked him for an interview, which gave him his first experience with American journalism. When he later saw the printed copy, he said the reporter had turned his comments "upside down."[10]

From Omaha, Plunkett took another detour from his journey to the ranch in order to visit the Close Colony near Le Mars, Iowa, which was managed by a Plunkett relative, Reynolds Moreton, a retired Royal Navy captain. This visit followed up on an earlier call Plunkett had made in England on the Close brothers, who owned the colony. When he met with the Close brothers, they talked about speculation in Minnesota land, just across the state line from Iowa. Plunkett's initial interest in the Iowa location is unclear, but he seems to have been considering the possibility of bringing western cattle to Iowa to fatten them up on corn in order to increase their value.[11]

The Close Colony consisted of a farm of 950 acres of rich soil, where sixteen pupils from the families of British aristocrats (Moreton's "pups") each paid £120 per year for the privilege of working on the farm. Plunkett saw the barefoot Lord Hobart, heir to the earldom of Buckinghamshire (though not the money), following a plow and commented, "I never saw anyone more like a ploughboy & less like an Earl." Then he added the telling imperative from his own background: "But the blood is there."[12]

---

[10] Ibid., 15–17 and 31 May 1881.

[11] Reynolds Moreton was related to Plunkett through the Duttons, on his mother's side of the family. Ibid., 24 Jan. 1881. Plunkett did join the board of First American Land Company, a Minnesota land speculation promoted by John Swatman. Ibid., 24 Jan. 1881 and 3 Feb. 1881. The Close brothers were William Brooks Close (born 1853), James Brooks Close (born 1851), and Frederic Brooks Close (born 1854).

[12] Ibid., 15–17 May 1881. Sidney Carr Hobart Hampden, born in 1860, whose courtesy title was Lord Hobart, was the grandson of the earl of Buckinghamshire. Back in England, his calling cards read, "Lord Hobart, Dealer in Hogs." After he succeeded to the earldom in 1885, the *Le Mars Sentinel* wrote, "He could not fail to take the prize for being the homeliest man in any contest either at home or abroad, but a good fellow for all that." Harnack, *Gentlemen on the Prairie*, 7, 112.

Plunkett was impressed with the Iowa land, saying, "I wish I had speculated at Le Mars rather than Powder River first. But then there is no romance except ploughing, &c, and it is not as healthy." After reaching Cheyenne, he succumbed to the temptation to invest in Iowa and wrote his father to ask for an advance of £1,200 to buy a farm from Reynolds Moreton. However, when he went back to Le Mars in October to have a closer look, his original enthusiasm evaporated. He found that the pupils were "idling away their time" and concluded that hopes for the colony were "low." This forecast later proved accurate, for by 1885, the pupil system of the colony was discredited, and few paying boarders were coming to the colony.[13]

Back in Omaha, Plunkett was still not ready to go directly to Wyoming, stopping instead at Grand Island, Nebraska. He wanted to visit an Irish colony located north of Grand Island in Greeley County, Nebraska, but he suddenly realized that journey would take too long. He needed to be in Cheyenne for the trial of the Searight case, so he pressed on west, and on May 20, two weeks after leaving New York, Plunkett finally reached Cheyenne. He was in Wyoming Territory but still a long way from the ranch.[14]

He took up residence in the Cheyenne Club (he approved of the accommodations), where he saw Gilbert Searight, who glowered at him. More disturbing was the gossip among others in the club, who said that a jury would likely vote against the British. The next day, Plunkett conferred with his lawyers and was very unhappy with their work, despairingly confiding to his diary, "The Law's Delay! Did Hamlet foresee Roche & Plunkett *vs.* Searight?" Nor did the situation improve on May 23, when the long-awaited hearing of the Searights came up before Chief Justice James B. Sener, who was sitting as Cheyenne district judge.[15]

---

[13] Plunkett diaries, 17 and 23 May 1881, 4 Oct. 1881, and 6 May 1883.

[14] The Irish colony Plunkett wanted to visit was apparently one established by Gen. John J. O'Neill, an eccentric Irish leader who had participated in the Fenian invasion of Canada and was imprisoned for his part in those matters. The O'Neill colonies were in Atkinson and Greeley Counties. The following year, Plunkett met at Avoca, Iowa, with a priest from the Catholic Colonization Society, which had a colony in that area. Olson, *History of Nebraska*, 173; and Plunkett diaries, 17 May 1881.

[15] Plunkett diaries, 20 May 1881. One of the benefits of the Cheyenne Club over the hotels was the fact that the local newspapers, which listed new arrivals at the hotels, did not print the club's guest list unless the guests wanted their presence known. Also, Plunkett diaries, 21 May 1881.

It was a short hearing, for the judge granted the Searight motion to strike Plunkett's claim for lost profits on the undelivered cattle and gave Plunkett & Roche ten days to amend their petition—if they could. Later in the summer, Horace vented his frustration over his treatment at the hands of the local leaders. Speaking of these men, who were socially as lofty in their society as one could be in that part of the world, he put his own words in their mouths, writing, "You have a social position. We have hardly any." He added, "They don't like me naturally & on the whole I don't like them."[16]

On May 25, Plunkett conferred with both Brown & Potter partners, meeting Melville C. Brown for the first time. Brown had served in both the Idaho and Wyoming legislatures and was the U.S. attorney for Wyoming from 1877 to 1881. "He's no fool," Plunkett said, and the conversation with Brown gave him hope that he might prevail in the case "at some future time." Unfortunately, nothing further would happen in the case until the fall term of the court, so Plunkett left Cheyenne the next day, this time bound for Powder River.[17]

While in Cheyenne, Horace took advantage of an opportunity to talk to one of the early ranchers in the territory, Nathaniel Russell Davis. Davis came to Wyoming in 1870, started a ranch south of Cheyenne, and by 1877 was reputed to be the largest cattle rancher in Weld County, Colorado. In an eerily prophetic comment, Davis told Plunkett that overcrowding on the public range would eventually bring about a catastrophic winter death loss. In just over six years, that prediction came true.[18]

On May 28, after thirty-seven hours on the stage, Horace Plunkett finally arrived at the ranch on the Powder River, at last ready to start the year's work. Boughton and the Roches had preceded him to the ranch, and he was relieved to find that they were not quarreling. Two days later, he saw foreman Jack Donahue in action at the roundup, which was then in progress, and declared himself "very well satisfied" with the man, an evaluation seldom

---

[16] Ibid., 21 July 1881.

[17] Ibid., 25–26 May 1881. Brown, who was then 43, was born in Maine and went to California in 1852 before going to Boise, Idaho, where he studied law. He came to Cheyenne in the fall of 1867 when the railroad laid out the town, and the following year he followed the railroad across the mountain to settle in Laramie.

[18] Ibid., 25 May 1881. Also, Woods, *Wyoming Biographies*, 75.

earned from Plunkett. To emphasize his high opinion of Donahue, he rated him as the best foreman in the region. All of these assessments were subject to change, as we shall see.[19]

Because the ranch did not have a cook, Edmund Roche, who at age twenty-two was the youngest of the five Roche brothers, was pressed into that duty. When Edmund went to the field to help with the hay, Horace, who was at the headquarters because he was feeling unwell, took over the varied domestic chores. Aside from cooking, there were four cows to milk, chickens to feed, butter to churn, and a foal and two young elk to be fed with milk (these being pets the partners had brought to the ranch house), wood to chop, a garden to water, and general cleaning work. Sleeping was also a problem, for when there was no bed for him, Horace had to sleep with Alexis, "& so had no sleep."[20]

In June, Alexis and Horace rode to Fort McKinney to attend a meeting to organize a local stockgrowers' association, and on the way they sampled the customary hospitality offered to travelers on the range. At the hay farm of one Chapin's ranch, they partook of free food and shelter, Plunkett remarking, "We should only insult him by offering money." The journey of 60 miles over rough country proved a waste of time, for there was no agenda for the meeting, and nothing was accomplished.[21]

Plunkett's decision to hire Walter Bulwer's son William also proved a bad choice. Initially, he had sized up young Bulwer as "not vibrant . . . but good stuff for our company," but in June, he sent Bulwer on an errand to the Frewens' ranch forty miles away, and on the way Bulwer forgot why he was making the journey. Plunkett then had to make the eighty-mile round-trip ride himself.[22]

In July 1881, Horace decided to try his hand at brokering land speculation deals to his friends in England. He wrote to Hugh Cullen to suggest that he represent Cullen on land deals in America, charging 5 percent of the purchase money and 10 percent of the profit on any sales of land. Cullen soon responded, saying he had asked Bishop O'Connor to buy £2,000 worth of Omaha lots for

[19] Plunkett diaries, 28–30 May 1881.

[20] Ibid., 13 and 15 June 1881.

[21] Ibid., 20–21 June 1881.

[22] Ibid., 19 March 1881 and 23 June 1881.

him and offering Horace a finder's fee; Plunkett replied that he would not take a fee on this first purchase but would charge one on future deals.[23]

The severe winter that had just ended convinced Plunkett that their range could soon be overcrowded, and as early as June 1881, he proposed to Alexis that they establish a corn feeding scheme, whereby cattle could be moved from the range to a facility where they could be fattened for market. Plunkett had apparently abandoned the idea of feeding in Iowa, and now he suggested they locate the operation on the Northwestern railroad in Dakota Territory. Later, after talking with John T. Stewart, who was ranching on the Sweetwater in Wyoming, he went to see Stewart's large feed and farming operation in Iowa. This trip caused Plunkett to change his mind again and decide Iowa really was a better location than Dakota.[24]

Plunkett saw the feeding scheme as a safety valve against overcrowding and also as a way to enter a new business: the livestock commission business. Not incidentally, the new location away from the ranch would provide an excuse for him to have Alexis manage the feeding operation and get him off the range. Plunkett would then continue managing the ranch on the Powder River, making only occasional visits to the feeding farm to oversee Alexis. Considering the frenetic schedule he was then pursuing, it is hard to see how he would have time to do everything this plan called for.[25]

At the spring roundup, the number of calves branded was an initial indicator of the health of the herd, but it was the fall roundup, which began at the end of August, that disclosed how many steers could be sent to market. Horace learned that the cowboys on the roundup found only 335 steers ready to market, when he had expected there would be 540. There was no way to know the reason for the deficiency: some may have been missed by the roundup but would be gathered in later roundups, whereas others may have died or been stolen. It was easy to speculate as to how many animals fell in each category, but the reality of the open range system was that only the ones gathered could be sold.[26]

---

[23] Ibid., 28 July 1881.

[24] Ibid., 5 July 1881. Stewart owned the 71 Quarter Circle ranch.

[25] Ibid., 25 May 1881 and 3 June 1881.

[26] Ibid., 28 Aug. 1881. During the fall roundup, the calves missed in the spring roundup were branded.

Plunkett sent the cowhands with the beef herd on the long journey south to the railroad and then rode on ahead to Point of Rocks, Wyoming Territory, where he stopped for the night. There he became "redolent of skunk" when a skunk climbed through an open window and startled the cook, who dispatched the poor animal with his shotgun in the room where Plunkett was sleeping. News of another possible speculation was waiting at the Union Pacific's Rock Creek station, where Horace met Dick Frewen, who had an elaborate idea of how to make money.[27]

Dick Frewen wanted to establish a stage line to bring tourists to Yellowstone National Park from the Utah Northern railroad, build hotels to house these tourists in the park, and buy carriages to move them to the various attractions. Plunkett liked the idea, although he knew the Frewen brothers would be difficult partners. He hoped it would be possible to obtain a monopoly on the service from the government, and that the railroads would also provide help. Plunkett immediately broached the Yellowstone idea to J. T. Clarke, superintendent of the Union Pacific, who said the railroad would consider the idea. Next Plunkett talked with banker Morton E. Post and Wyoming governor John W. Hoyt. Later, he met with a Union Pacific official in Council Bluffs, Iowa, who flattered him by saying the railroad wanted to support the Yellowstone scheme because of the "influential" British involved. In the end, Lord Dunsany opposed the idea and nothing came of it, but in the meantime Horace had consumed much of his scarce time and energy.[28]

Plunkett's health was troubling him again, and before going east again, he stopped in Laramie to consult with Dr. James H. Hayford, whom Plunkett described as "a man not of genius, but of great pains." This description seems too harsh for a man who was a teacher, went to medical school, and studied and practiced law before coming to the Wyoming country in 1867, before the territory was established. In Cheyenne, Hayford took up journalism then moved to Laramie, where he served as editor of the *Sentinel* while also serving as postmaster and justice of the peace. When Plunkett came to see him, Hayford was fifty-four and still practicing medi-

[27] Ibid., 7 Sept. 1881.

[28] Ibid., 8, 9, 13, 26 Sept. 1881 and 23 Oct. 1881.

cine. Plunkett gave the doctor the following summary of his symptoms: "The nervous weakness & excitability wh[ich] led to uneven temperament & inconsistency of character & the fear of hysterical children in case of marriage & general effects of weakness on character." After hearing this summary the good doctor prescribed dilute phosphoric acid for the physical symptoms and for the psychological complaints advised, "Don't think about it." Plunkett mused that these prescriptions were "no doubt right.[29]

Plunkett followed the steers to Pine Bluffs, Wyoming Territory, where they were to be loaded on cattle cars. When he got there, he found that the herd had been fenced all night without feed and was not ready to load. Finally the steers were loaded, twenty to a cattle car; they had to be prodded periodically to prevent suffocation. When the train stopped, the men accompanying the cattle walked alongside the cars to check on the animals (carrying lanterns at night), and when the train started the men climbed on top of the cars or rode in the caboose. In these straitened circumstances, Plunkett made the two-day trip with the cattle to Council Bluffs, Iowa. The herd suffered considerably during the trip, which contributed to the "miserable" prices they brought when they were sold.[30]

Back in Cheyenne on November 7, the motions on the Searight case came before the fall term of the court, and this time Brown represented Plunkett & Roche. Plunkett thought Brown made a stronger impression than Corlett, although he conceded that Corlett scored some points as well. After the oral arguments, the judge again reserved judgment. Because Plunkett was once again preparing to depart for Ireland for the winter, he arranged to give a deposition of his testimony so that the case could go forward without the need for him to appear. Before Plunkett left for the East,

[29] Later in the evening after seeing Hayford, Plunkett discussed religion with some of the cattlemen, which led to a "keen intellectual fight," resulting in laurels both to agnosticism and Roman Catholicism. Plunkett diaries, 14 Sept. 1881. Hayford was born in Potsdam, New York, on 26 Dec. 1826, and I have given only an abbreviated list of his astonishing experiences. He taught school in New York, Iowa, and Illinois; attended the University of Michigan medical school; and studied law in Wisconsin, where he practiced for a time before going to Colorado. He went to Cheyenne in 1867 when it was still a tent city, worked on the *Cheyenne Leader,* and bought the *Laramie Sentinel* in 1870. Keen, "Wyoming's Frontier Newspapers."

[30] Plunkett diaries, 23–30 Sept. 1881.

Brown told him the case would not be decided before May of 1882.[31]

Plunkett had spent just three months of 1881 on the range, and the partnership operation of Plunkett & Roche was not well run, not only because of Plunkett's absence, but also because of the shortcomings of his partners. Alexis Roche did not fill the role of managing the ranch while Plunkett was away on necessary travel. Plunkett had hoped that Boughton could look after matters around the headquarters, including record keeping, but Boughton did not learn proper accounting and was liable to forget to feed the chickens and milk the cows.[32]

Moreover, for the British ranchers, foreign visitors added work that their American counterparts did not have. The Powder River ranches attracted many guests from abroad in 1881, and although much of the entertainment responsibility fell to the Frewens, Plunkett also knew many of the guests and had to bear a part of the burden they represented. The parade of visitors began in June with Captain Edward E. Shearburn, a hero from the Afghan conflicts. Unfortunately, Shearburn had lost his money gambling and had come to the Powder River to try his hand at ranching. In Wyoming, he suffered a bad fall from a horse and was in the Fort McKinney hospital for weeks, and although he was still eager to go into the ranching business, Plunkett dismissed the idea, saying they had no need for "authoritative inexperience."[33]

In August, Lord Mayo arrived, and because he was a Plunkett family connection, Horace had a certain responsibility for him. Soon Mayo was causing trouble with the locals. By the time he returned home in October, Mayo had fallen from a horse and broken a bone in his hand. When Mayo left the ranch, Horace asked the earl to pay for some of the service he had received (some $70

---

[31] On the way to New York, Plunkett's passenger train collided with a freight train while he was in the diner, and two days later he was involved in another railroad accident when the Pullman car derailed. These minor train wrecks were not uncommon, and the Cheyenne newspapers often carried no report of them. Plunkett diaries, 7 and 18–26 Nov. 1881.

[32] Ibid., 26 June 1881; 29, 30, and 31 July 1881; and 26 Aug. 1881.

[33] Shearburn led 17 charges by 120 men against a force of 10,000, escaping with bullet-riddled clothing and only 42 surviving soldiers. Plunkett diaries, 8 June 1881. Shearburn later became a partner of the Frewens in their venture to sell bat guano from Texas caves. Woods, *Moreton Frewen's Western Adventures*, 47.

less than Plunkett had spent), and Mayo complained about the cost.[34]

In October, a number of Plunkett's friends came to the Powder River ranches. The list included Gilbert "Gillie" Leigh, who later lost his life in the Big Horn Mountains; Lord Manners and his brother, Claud; Bernard "Barnie" Fitz-Patrick; Lord George Montagu; and Lord Donoughmore. All of these gentlemen had their own specific needs—and idiosyncrasies—requiring much effort to satisfy them (not always successfully). Not incidentally, all of that effort detracted from the business of cattle ranching, but Plunkett nevertheless concluded, "They are welcome in our solitude."[35]

After Plunkett and his partners left the ranch for the winter, the men who were not needed for winter work were laid off and left to fend for themselves, riding the grubline at the ranches for free food and shelter. Jack Donahue, the first foreman, and a skeleton crew of men stayed at the Plunkett & Roche ranch house, which was finally chinked so that it was warm. The cattle out on the range received little attention; they ate the cured grass when it was not covered by snow. Winter storms scattered them as they sought shelter in the gulches, and in the spring roundup the cowboys would have to try to find them and bring them back to their customary range.[36]

Plunkett sailed for home from New York on November 26, 1881, and he turned his attention to raising more money for the following year. To Hugh Cullen, he proposed that Cullen invest £5,000–£10,000 and give Plunkett the management responsibility, which brought forth a Cullen investment—but only for £1,000—at the beginning of 1882. Plunkett was more successful with Thomas Leonard, who invested £4,000 on the same terms. In March, Barnie Fitz-Patrick loaned Plunkett £2,000, and Lord Mayo advanced $3,000.[37]

---

[34] Plunkett diaries, 11, 14, and 31 Jan. 1881; 27 Aug. 1881; and 16 and 25 Oct. 1881. Lady Mayo was descended from the Earls of Bessborough, as was Chambre Brabazon Ponsonby, who married Horace's sister, Mary Sophia Eliza Plunkett.

[35] Ibid., 11 Oct. 1881. Gillie Leigh, heir to Lord Leigh, was a member of Parliament; Bernard Edward Barnaby Fitz-Patrick was the heir of Baron Castletown; and Lord George Francis Montagu was the younger son of the Duke of Manchester, who was chairman of Frewen's Powder River Cattle Co.

[36] Ibid., 16 Oct. 1881.

[37] The Fitz-Patrick and Mayo deals were for specific cattle purchases, and in May 1882 Plunkett wrote Fitz-Patrick that his investment would be diverted to stock in the new corporation, rather than for a specific herd. Ibid., 16 Dec. 1881, 11 and 14 Jan. 1882, 15 May 1882, and 19 Dec. 1883.

To get help with some of his American entanglements, Plunkett made a concerted effort to persuade a close friend from Ireland, Cyril Coleridge, to join him in Wyoming. Coleridge was a sporting friend who bought horses for Plunkett, and Plunkett suggested they become involved as land brokers. Being a land broker would have given Coleridge plenty of flexibility to take off time from Wyoming business and return to Ireland from time to time. Unfortunately, Coleridge would not agree to come.[38]

So ended 1881 for Horace Plunkett, a year in which he entertained many ideas without much result; traveled many miles, often in considerable discomfort; and aroused the hostility of the locals by starting a lawsuit that he would soon learn he could not win. Withal, it was not a good year.

---

[38] Ibid., 12 July 1881, and 11 and 12 Dec., 1881.

4

Corporations and More Partnerships

After Plunkett left Wyoming Territory for home in November 1881, the Searight lawsuit continued its fitful course in Cheyenne. At the end of November, Justice Sener again struck down Plunkett's request for lost profits from the undelivered cattle. At the beginning of December, the Searights said they had a written contract with Plunkett & Roche, in which Plunkett & Roche agreed to accept the cattle that were actually delivered on November 3, 1880. It is likely that Corlett had some writing to base this assertion on, and Brown & Potter asked the Searights to produce a copy of this contract. However, on December 30, the judge struck the Searight claim regarding the additional contract and did not require them to produce it. After more pleadings in January and February 1882, the case once more was deferred to the next term of court.[1]

While this was going on, a very bad omen for the Plunkett & Roche side of the case arose outside the courtroom rather than from the pleadings before the judge. There was a strange absence of press coverage of the dispute. When the court was in session, the local papers covered court proceedings every day, sometimes in considerable detail, but they ignored the Searight case. Indeed, the case was not even listed as one of those being heard by the judge. Obviously, this controversy was not popular in Cheyenne.

[1] One can only speculate what the additional "contract" might have been, because Plunkett did not mention it in his diaries. Perhaps Alexis gave the Searights some paper at the time the cattle were delivered to Plunkett & Roche.

Plunkett landed in New York on May 1, 1882, where he pursued more land speculation ideas before arriving in Cheyenne on May 21 on his way to the ranch. At the ranch, he began buying cattle to replace the deficiency from the Searight contract, finally purchasing two herds. He soon needed more money, and at the beginning of August he asked for another £3,000 from Lord Dunsany. With this resource in prospect, he then bought a one-third interest in a partnership with Henry J. Windsor and John Coble. Following his usual practice, Plunkett wrote the agreement for the Windsor, Coble & Plunkett partnership, which continued for a number of years.[2]

Early in July, Plunkett took delivery of 1,450 head of cattle at Rock Creek, where Jack Donahue and Nate Champion met him, bringing forty-two horses and a crew from the ranch. Plunkett and the crew then spent ten days rebranding the cattle before starting them north to the ranch. Nate Champion, who was not yet twenty-five, later lost his life in the so-called Johnson County War of 1892. He was born in Texas, where his Standifer relatives were prominent in Milam County. He may have been attracted to Wyoming because of the experiences of a distant cousin, Captain Jefferson Standifer, who first came to Wyoming Territory in 1866 and died at Fort Fred Steele in 1874. Early reports described Nate Champion as good-looking, brave, honest, forthright, and handy with a gun, and he had many friends among cowboys and small ranchers. Although Plunkett later laid him off, it was because he had no work for Champion, not because Champion had any deficiencies as a cowhand.[3]

---

[2] The agreement with Windsor and Coble was signed on 11 Sept. 1882. Plunkett diaries, 3–9 July 1882; 1, 4, 5, 7, and 11 Aug. 1882; and 11 Sept. 1882. The Windsor, Coble & Plunkett partnership had an initial capital of $93,000, of which $31,000 was the cash Plunkett paid for his one-third interest. The cattle on the range from the Windsor & Coble partnership were valued at $86,000, and the new partnership also assumed nearly $30,000 in Windsor & Coble debt. All of the debt was refinanced during the first year by substituting new borrowings, including a total of $5,500 from Plunkett. The Windsor, Coble & Plunkett journal is in the possession of Fred Hesse, Buffalo, Wyo. Henry J. Windsor was born in Maryland in August 1855, and John C. Coble was born in Pennsylvania in June 1858. Like Plunkett, neither was married when their partnership was formed.

[3] Plunkett received 700 head from Fillmore and 750 head from Fisher at Rock Creek on 4 July. Plunkett diaries, 24 and 27 June 1882 and 4 July 1882. Nathan David Champion was born in Williamson County, Tex., on 29 Sept. 1857. After his mother's death in 1864, he went to live with his uncle, David Standifer. Three of Champion's brothers also came to Wyoming Territory. His older brother William was ranching at Mayoworth in Johnson County before 1879, younger brother Dudley was shot the year after Nate died, and another younger brother, Benjamin, was also in Johnson County in 1900.

Plunkett was not the only British rancher looking for expansion opportunities in 1882: in August, Moreton Frewen incorporated his ranch as the Powder River Cattle Company, Ltd., and he was casting about for ways to spend the money his new shareholders (who included Lord Dunsany) had given him. He offered to buy Plunkett's EK herd for $175,000, but Plunkett countered with a price of $260,000, and Frewen told his wife the counteroffer was "simply crazy." Still, Frewen was determined to make a deal with Plunkett, and in October he agreed to buy all the next year's beef cattle under the Plunkett & Roche EK brand for 5¢ a pound, delivered to Chicago. This contract, which proved to be a foolish bet on the market by Frewen, later caused a good bit of difficulty between the two men.[4]

Plunkett had decided in the fall of 1881 that he would fence a portion of the range to improve control over their cattle, and this fenced pasture was ready for use in the fall of 1882. It was large enough to accommodate all the beef gathered for market in the fall roundup, a total of nearly 650 head, plus 150 head from Plunkett's new partnership with Windsor & Coble. Unfortunately, fencing on the public domain was illegal, precisely because everyone was theoretically entitled to use all of the range. Still, the practice continued, and ranchers built many miles of illegal fences on the public domain.[5]

On September 26, Plunkett ordered thirty-three rail cars to be at Rock Creek on October 1 to ship their beef cattle to market, effectively providing for 660 beef animals. Early in October, Horace sold ninety head in Omaha and the remainder in Chicago two days later, in what Plunkett called a "flat" market. The Omaha steers averaged over 1,100 pounds, yielding over $43 per animal.[6]

Plunkett now took off a few days for sightseeing in the Rocky Mountains with Boughton and other friends from home. They first went to Denver (which had grown considerably since Horace and

[4] Plunkett received Frewen's offer from Frank Kemp on 18 Aug. 1882 and wrote his counteroffer to Frewen in Cheyenne the same day. Moreton Frewen to Clara Frewen, 8 Sept. 1882, Frewen collection. Also, Plunkett diaries, 18 Aug. 1882. Shortly after Frewen made his offer to Plunkett, the latter sold steers in Chicago for 3.75¢ per pound, so the Frewen offer was a very strong bet on the following year's market.

[5] The 1882 fall beef tally was 638 head. Plunkett diaries, 29 Oct. 1881 and 4, 9, and 11 Sept. 1882.

[6] Ibid., 26 Sept. 1882 and 1, 6, and 9 Oct. 1882.

Gussie Briscoe were there in 1879) and from there to Manitou Park and to the Garden of the Gods. Leaving the others, Plunkett and Boughton went to Leadville, where they inspected an iron mine.[7]

Early in November, Horace went to Laramie, where the Searight trial was finally to be held. Horace was convinced that Justice Sener was discriminating against him, and in June he finally lost patience and asked for a change of venue from the Cheyenne court. This request was granted, and the case was then given to Jacob B. Blair, another Wyoming Supreme Court justice, who was hearing cases across the mountain in Laramie.[8]

In Laramie, Plunkett dined with Bill Nye, the witty editor of the *Laramie Boomerang*. Nye was also the Laramie postmaster, and when Assistant Postmaster General Frank Hatton (also a former journalist) notified Nye of his appointment, Nye did not miss the chance to have fun with official Washington. "Now that we are co-workers in the same department," he wrote, "I trust that you will not feel shy or backward in consulting me at any time relative to matters concerning postoffice affairs." Hatton showed the letter to President Arthur, who also enjoyed it.[9]

Nye and Plunkett became friends, and when Nye wrote his play *The Cadi*, which enjoyed a run of 125 days in New York, one of the characters was patterned after Horace Plunkett. The character of Taylor Wellington is described in the script as "Wild, young son of a good Irish family. He gets busted and deals faro, ashamed to go home or write. Falls heir to fortune and title." The caricature was surely meant in fun, and Horace obviously accepted it as such.[10]

Finally, in November 1882 Plunkett got the chance to have a jury hear his side of the Searight case. He testified for an entire day in front of a jury composed mostly of Laramie residents. However,

---

[7] Ibid., 20–29 Oct. 1882.

[8] The case of *Roche and Plunkett v. Searight Brothers* was given Case No. 1111 in the Second Judicial District (Albany County). Also, Plunkett diaries, 27 June 1882 and 13 and 14 Nov. 1882; and the *Cheyenne Daily Sun*, 16 Nov. 1882. Justice Jacob B. Blair, who also sat as district judge for the Second Judicial District, was born in Virginia (later West Virginia) on 11 April 1821 and served in the U.S. House of Representatives from Virginia and West Virginia before coming to Wyoming in 1876.

[9] Edgar Wilson "Bill" Nye was appointed postmaster in the summer of 1882, and his official acceptance was dated 9 Aug. 1882. Nye, *Bill Nye*, 98, 308. Also, Plunkett diaries, 9 Nov. 1882.

[10] Nye's play *The Cadi* was originally titled *The Village Postmaster* but did not do well in the small towns of Illinois. The play was revised somewhat for its run at the Union Square Theater in New York. Nye, *Bill Nye*, 102, 305.

after suffering under Corlett's grilling, Horace realized he was not going to win the case and agreed to a compromise settlement. He did not describe the settlement in his diary, but it is clear that the locals prevailed against the British gentlemen.

Plunkett blamed his failure to recover against the Searights on local prejudice and the incompetence of his lawyers, but it is not evident that either factor was decisive. The heart of his claim was the loss of $6 per head on the 2,000 head of cattle the Searights didn't deliver. As it happened, it would have been impossible to prove that Plunkett & Roche was actually injured by the Searights' failure to deliver 2,000 head of cattle. A local jury aware of the cattle losses in the 1880–81 winter might even wonder if the Searights had not actually conferred a favor on Plunkett & Roche, for that partnership did not lose any of the 2,000 head in the harsh winter.

Plunkett continued his efforts to incorporate his ranching operations, hoping to use the new corporation as a vehicle for future expansion, where he could consolidate management in his own hands. He did not have the money to buy out his partners, and he immediately offered Alexis a share in the company. As to Boughton, Plunkett's feelings were more complex, for he wanted to keep Boughton's money in the corporation in order to avoid the expense of buying him out, but at the same time Plunkett didn't want to pay Boughton a salary. To his relief, Boughton promised not to interfere with management of the corporation and to accept only a "nominal" salary, and on this basis he joined the Frontier Cattle Company in the summer of 1884.[11]

Plunkett made a considerable effort to attract other ranchers to his new corporation. He had many conversations with Peters and Alston, but Alston finally decided not to join. Plunkett also approached Henry Augustus Blair, who owned the Hoe Ranch, and Blair came to look over the Plunkett & Roche operation—which had now been folded into the corporation called Frontier Cattle Company—but decided not to join the new company.[12]

When he did not succeed in expanding his operation by mergers with other ranches, Plunkett decided to buy more cattle. Windsor

---

[11] Plunkett diaries, 16 and 17 Oct. 1881 and 11 July 1884. It is unknown what Boughton's investment in the partnership was, but he later said that he owned $65,000 worth of Frontier Co. stock.

[12] The Hoe ranch headquarters were near the forks of the Powder River. Ibid., 27 Aug. 1882, 27 Nov. 1882, 16 Jan. 1883, 30 and 31 May 1884, 19 June 1884, and 13 July 1884.

told Plunkett that Worden P. Noble's herd in the Big Horn Basin was for sale, and Plunkett cast about for another investor to share the cost. Back home in England at the end of 1882, he went to see Algernon James Winn (another Plunkett family connection), who wanted to join in the Wyoming cattle ranching business. Winn agreed to put up one-third of the cost of the Noble herd, and the partners agreed to buy the herd for $153,000, on easy terms. Winn still had to raise his share of the purchase price, and he set about to do that.[13]

The first year's results for the Windsor, Coble & Plunkett partnership showed a profit of $48,000, but this total concealed important information, for that sum included an estimate of nearly $57,000 representing the unrealized increase in the value of the cattle still out on the range. This estimate valued the natural increase in numbers as well as the gain in weight during the year. Thus, the entire profit for the year was still unrealized and was subject to the vagaries of life on the range as well as future cattle markets. Because the book profits could not be used to buy anything, Plunkett's most urgent task still was to raise more money. Early in January 1883, he raised $30,000 from his friend William Blacker, and in April Tom Leonard advanced another $5,000 in a deal entitling him to a quarter of the profits arising from that investment.[14]

After these busy times in England and Ireland, it was time for Plunkett to return to America, and on April 20, 1883, he sailed from Queenstown for New York. As usual, after landing he did not go directly to Cheyenne. As he done two years before, he stopped in Le Mars, Iowa, to see Reynolds Moreton, who now proposed to sell his farm for a cattle-feeding facility. Then in Omaha, Nebraska, Plunkett saw Bishop O'Connor again, and the two men discussed Omaha real estate. He got to Cheyenne by May 8, where Winn met him with the bad news that he couldn't raise his share of the money for the Noble WP herd.[15]

Lacking the resources to make the entire WP purchase, Plunkett went to see Tom Sturgis of the Union Cattle Company to negotiate a

[13] Ibid., 12 and 26 Dec. 1882, 15 Jan. 1883, 8 May 1883, and 9 Sept. 1883. Algernon James Winn was a younger son of Rowland Winn, the future Lord St. Oswald, and his sister was married to Valentine Lawless, fourth Baron Cloncurry. Horace Plunkett's grandmother was the daughter of the first Baron Cloncurry.

[14] Ibid., 15 and 26 Jan. 1883.

[15] Ibid., 19 and 20 April 1883 and 4–8 May 1883.

quick resale of the herd. Tom Sturgis, the secretary of the Wyoming Stock Growers Association, was a powerful figure in Wyoming Territory, and Plunkett probably was aware of the Sturgis family before coming to America. The London branch of the Sturgis family, headed by Russell Sturgis, was distantly related to the New York branch of the family, and both were financially prominent. Russell Sturgis's three sons all attended Eton, and the youngest, who was in Plunkett's class, was very aware of his American cousins. Tom Sturgis readily agreed to buy the WP herd for $200,000, paying for the cattle with Union Cattle Company notes, bearing interest at 10 percent. The price was attractive, and the profit was welcome—if the notes were paid when they fell due. Unfortunately, this transaction did not end Plunkett's association with the WP herd.[16]

The deal with Sturgis was a part of Plunkett's new strategy for conducting business in Wyoming, for in the wake of the Searight debacle, he did not want to be on the losing side again. He now set about making business arrangements with leaders in the territory. Indeed, the day after he compromised with the Searights in the fall of 1882, he agreed to join the Cheyenne group that was organizing the Wyoming Development Company. This company was planning the Wheatland Project, a huge irrigation development in the northern part of Laramie County. The organizers of the company consisted of former governor John W. Hoyt, future governor and senator Francis E. Warren, former justice and future senator and governor Joseph M. Carey, and congressional delegate and banker Morton E. Post. Plunkett became an important member of this blue ribbon group.[17]

The Wheatland Project was different in more than one way from Horace Plunkett's other early ventures in the West. His ven-

---

[16] Russell Sturgis, head of the London branch of the family, was the senior partner of Baring Brothers. In New York, Tom Sturgis's brother Frank was a broker and governor of the New York Stock Exchange and later became the exchange president. Both branches of the Sturgis family descended from Edward Sturgis, who came to the Plymouth Colony in 1640. Russell Sturgis's youngest son, Howard Overing Sturgis, born in 1855, was at Eton with Plunkett, although he did not live in Plunkett's house. Howard Sturgis is known for two novels, one of which, *Tim*, was the story of love between two Eton boys. Benson, *Memories and Friends*, 266, 268. Also, Plunkett diaries, 22 Oct. 1883.

[17] Plunkett diaries, 17 Nov. 1882. The Wheatland Project shareholders were clearly leaders in territorial affairs, and there are the usual assertions that their political influence played a part in the state engineer's favorable adjudication of the Wheatland priorities for irrigation from the Laramie River.

ture in open range ranching did not involve large initial investments, but the Wheatland Project contemplated heavy investment to bring water to the land before any of the land could be settled. Early on, Plunkett recognized that a major weakness in his ranching venture was the inability to keep others from grazing the free grass his cattle were living on. Although he fenced a pasture, this was illegal and did not overcome the overcrowding risk. In the Wheatland Project, value was to be created by bringing the scarce water from streams to irrigate the land, and in this case the government sanctioned a monopoly over the water by appropriating the water right to this land. For Plunkett, this feature overcame the disadvantage of the high initial investment required.

Plunkett's management relationship to the Wheatland Project was also very different from the ranching business, where he always insisted on being recognized as the key decision maker. In the Wyoming Development Company, he was never the majority shareholder, nor was he ever designated as the chief manager of the company. Nevertheless, despite this lack of formal recognition, he soon gained considerable—and increasing—influence over its planning and management. The project was extremely "long-winded," as Plunkett put it, for it was organized in 1883, did not open for general settlement until 1894, and was finally completely liquidated some years after Plunkett's death in 1932.

The Wheatland Project was conceived by John Gordon, an Irishman who came to the Greeley Colony in Colorado in 1870 and saw the possibility of developing a similar colony on the flat land northwest of John Hunton's ranch at Bordeaux. Judge Joseph M. Carey heard the idea from Gordon and had the project surveyed in 1881. The project was an ambitious scheme to develop a 58,000-acre tract located south of the Platte River in the northern part of Laramie County. The farms were to be irrigated with water from the Laramie River and from the Sybille Creek, and initially the land was expected to be acquired by settlers under the 640-acre Desert Land Act, although there were later some filings under the Carey Act.[18]

[18] Joseph M. Carey, a lawyer, was born in Delaware on 19 Jan. 1845 and in 1869 was appointed as the first U.S. attorney in Wyoming. Later, he was appointed as an associate justice of the Wyoming Territorial Supreme Court; he was elected Wyoming territorial delegate to Congress; and in 1890, he became senator for the newly admitted state. He was elected governor in 1910 and served until 1915; he died 5 Feb. 1924. The Wheatland Project also made use of the Carey Act, the terms of which became available to the state in 1895.

The laws covering the use of water for irrigation were in somewhat of a muddle in Wyoming at this time. In the eastern states, where irrigation was not common, the riparian system gave the right to use water to those on the stream banks. However, in the West, water for irrigation had to be conveyed some distance from the streams, requiring a system for appropriating the water to the land to be irrigated. Utah Territory pioneered an appropriation system for irrigation, and indeed, the first priorities for water use in Wyoming were apparently established under Utah control.[19]

Nevertheless, a strange 1875 Wyoming law purported to set up a sort of riparian system for the territory (the law also recognized all other systems then in place). Despite this confused legislation, as a practical matter, irrigation projects before 1886 (when the law was changed) were developed as though Wyoming was an appropriation jurisdiction. The first priorities for the Wheatland Project on the Laramie River were in 1883, and because the Laramie River soon proved unequal to the hopes of all the developers who had irrigation projects, that priority was of great importance.[20]

The Wyoming Development Company was formally incorporated in the fall of 1883, and the first shareholders were Post, Carey, Warren, Plunkett, William C. Irvine, and Andrew Gilchrist. Gilchrist, a Scotsman, joined the group in July 1883 on Plunkett's recommendation of him as a quality person (despite his lack of social background). Each of these six men subscribed equally to the initial $28,200 stock issue, and each would be called on to make many more investments in the following years. At first, Plunkett did not bring other British investors into the Wheatland Project, because

---

[19] In Wyoming, the right to use water was by "appropriation," according to the time of filing, which was its "priority." The first priorities in Wyoming were for 1862, in Uinta County, in the former Green River County, Utah. Although there is no evidence that an irrigation company was organized in Utah for these early ditches, the Utah law clearly contemplated that action, and the Wyoming priority indicates the earlier activity was recognized, whether under Utah law or under the federal laws of 1866 and 1870.

[20] The Wheatland Project appropriation (58,503 acres, 23 May 1883) effectively doomed the full development of the large project on the Laramie Plains (49,030 acres, 1 Oct. 1884) that was promoted by the Pioneer Canal Co. and its successor, the Wyoming Central Land and Improvement Co. The 1875 law has caused some to infer that Wyoming was a riparian jurisdiction before 1886, which was clearly not the case; after 1875, perhaps the territory was a mixed system, somewhat like the early experience of California. Although there were many priorities before 1886, it was the 1886 legislation that established the system by which the priorities were finally adjudicated.

he was concerned that management of the project was weak. Each of the other shareholders was deeply involved in other businesses, and he wondered if they would devote the necessary time to Wyoming Development Company.[21]

From the beginning, Plunkett exercised considerable influence over the project. At a meeting of the prospective shareholders in May 1883, Plunkett argued that the company should concentrate on selling land where it could to raise capital for the development work. At this meeting the board asked Gilchrist and Plunkett to inspect work on the project and report back to the board. When Gilchrist inspected the Wheatland Project, he was so discouraged that he wanted to leave the project; Plunkett only persuaded him to reconsider by promising him the job of superintendent. The board of the company later ratified this promise to Gilchrist, thus giving Plunkett and his friend effective control over the affairs of the Wyoming Development Company. By late fall, Plunkett expressed pleasure that the "sadly muddled" affairs of the company were being put in shape. When the company hired Gilchrist's Scottish friend John Chaplin to be the bookkeeper for the company, Plunkett and Gilchrist secured control over the accounting as well.[22]

---

[21] Governor Hoyt was a member of the original group but did not invest in the Wheatland Project. The *Wheatland Record-Times* asserted that Plunkett was sent to England to raise money. *Platte County Record-Times,* 2 April 1980. In the summer of 1884, Warren left the company and sold his shares to Thomas Sturgis. Sturgis proved an asset to the financing of the company, bringing access to investors in New York, the most important of whom was Robert Fulton Cutting, president of the Citizens' Union of New York, who remained a substantial shareholder in the company until its liquidation. Andrew Gilchrist, who was born in Scotland in 1844 and served in the Life Guards, came to the United States in 1865 and ranched in Greeley, Colo., before coming to Wyoming in 1877. According to Plunkett, Gilchrist had built up a fortune "from nothing." Plunkett diaries, 9, 11, and 18 May 1883; and 6, 9, 11, and 12 July 1883.

[22] Gilchrist's salary was $2,500 per year. John Chaplin, who was born in Scotland in about 1853, came to the United States in 1880 and was a clerk for the U.S. Geologic Survey in Denver before coming to work for the Wyoming Development Co.; he later became Plunkett's agent in Cheyenne. Plunkett diaries, 17 Nov. 1881, 10 May 1883, and 17 May 1884. Following the change in management Plunkett was sufficiently encouraged by the company's prospects to recommend the stock to his British friends. Gilchrist wanted to reduce his investment in the company, and Plunkett suggested that Henry P. Maxwell buy half of Gilchrist's stock. There were now seven shareholders, although the investment of Maxwell and Gilchrist was only half that of the others, proportionally reducing their obligation for subsequent assessments. Plunkett also sold 1,000 of his own shares to William Blacker and 300 shares to William D. Watson-Smyth. Plunkett diaries, 15–17 and 27 July 1883 and 24 Sept. 1883.

Plunkett and Gilchrist's influence over development of the Wheatland Project received virtually no public attention in Wyoming, and the local members of the board of the Wyoming Development Company continued to hold the spotlight in the public eye. Nevertheless, the practical effect of the board action in 1883 was that four leaders in Wyoming territorial affairs, ranging in age from thirty-eight to fifty-two, turned over the management of their company to an Irishman who was not yet twenty-eight and who had no real management credentials at all. This is remarkable, but it is even more remarkable that there is no evidence those local board members ever regretted that decision.[23]

Having thus reorganized the Wheatland Project, Horace left for the northern ranches on May 22, suffering the full measure of discomfort that commonly awaited travelers in that region. He was stuck on the road three times and finally had to walk with other passengers the rest of the night until they reached La Prele Creek, which was in flood. Using a rope stretched across the creek, four men and an "unhappy" woman with her child got across, and after spending the night outside in freezing weather they continued on the muddy road in a wagon. The horses soon played out, and the men again had to walk from Fort Fetterman to Sage Creek, where they got fresh horses, who also soon played out. The woman and child got a ride in a buggy, but the men walked another eight miles to Brown's Springs, which they reached at midnight. By then, Plunkett was lame, and he took the stage to Dry Cheyenne, finally reaching Frewen's Powder River store at 8:00 P.M. After enjoying a night's sleep in a bed, Plunkett made his way to his own ranch in a buggy Moreton Frewen sent for him.[24]

At his ranch, Plunkett found much that was not as he had left it. The ranch had a new Chinese cook, Yup Mi, and Plunkett wrote in his diary, "Hope the cowboys don't shoot him." The house was filthy, and the cowboys had treated the place badly, stolen many things, and "abused" the rest. After a day devoted to household chores, Horace prepared to join the spring roundup, which was already in progress on the other side of the Big Horn Mountains. Jim Winn was there for the trip as well, for after he gave up the idea of buying into the WP

[23] Hoyt was born in 1831, Post in 1840, Warren in 1844, and Carey in 1845. In the fall of 1883, the Wyoming Development board elected Plunkett a vice president of the company.

[24] Plunkett diaries, 23–26 May 1883.

herd, he had bought his own trail herd and established his own cattle operation, the Big Horn Cattle Company, over in the Big Horn Basin. Still another companion on the ride was Clement Finch, a younger son of Lord Aylesford who was visiting Winn.[25]

On the way over the mountain, the three men had Bob Stewart as a guide, and they slept in a cabin on Canyon Creek and took their noon meal at Harvey Booth's camp, following the etiquette of the range. Finch then left to return to Plunkett's ranch on the eastern side of the mountains, apparently taking Stewart with him, and Plunkett and Winn pressed on alone, at times in a drenching rain. Five days out, they crossed the raging Ten Sleep Creek, nearly drowning Plunkett's horse "Brownlow." Winn elected not to cross the creek, returning instead to his ranch on Canyon Creek.[26]

Needing someone to guide him to the roundup, Plunkett hired Frank Sykes (an irascible trapper who was living at the foot of the mountain) to guide him to Shell Creek, some twenty miles to the north. After a wet day's ride, Plunkett shared a bed with a "filthy" cowboy, and in the morning breakfast consisted of a stick of bad bread, some "worse" bacon, and a cup of water. Turning back to the south across thirty miles of badlands, they finally struck the trail of the roundup, which led to the mouth of the Nowood River. The river was running high and was twenty-five yards wide. Poor Brownlow had not fully recovered from his fearful crossing of Ten Sleep Creek, and on the way across the Nowood he gave up the struggle in midstream so that Plunkett was forced to swim to shore, leaving the hapless Brownlow to drown. Sykes salvaged Plunkett's saddle, two coats, and some important papers. Plunkett had to walk a further six miles before reaching the roundup.[27]

By June 14, Plunkett was back at his ranch, and the next day he went to see Willie Peters to propose a partnership among Peters,

---

[25] Ibid., 27 May 1883 and 4 and 5 June 1883. In May 1883, Moreton Frewen bought a trail herd for Jim Winn. Moreton Frewen's journal, 12 May 1883, Frewen collection. Clement Finch's brother was ranching in Texas, where he died in 1885. In the fall of 1886, Clement and his brother Daniel again came to visit Winn.

[26] Plunkett diaries, 6–10 June 1883. Plunkett often named his horses for some aspect of their appearance or behavior, and "Brownlow" undoubtedly referred to the horse's markings.

[27] Plunkett did not comment on the fact that Sykes, the guide, rode while Plunkett, his employer, trudged along on foot; however, those who knew the irascible Sykes would not have questioned the apparent anomaly. Ibid., 4–12 June 1883.

Plunkett, Boughton, and Tebbetts to start a hay ranch and horse-herding establishment north of the Plunkett & Roche ranches. The partners in this enterprise, which was called Tebbetts & Co., were Plunkett, T. W. Peters, Boughton, and Marston Tebbetts, with Tebbetts as the local manager. Tebbetts—whom Plunkett assessed in his usual style as "thick-headed, hard working"—was their neighbor on the range and was now down on his luck. Although Plunkett was only one of four partners and was not even listed in the partnership name, he was effectively the manager of the new organization. At the beginning of July, he sent Tebbetts to Dodge City, Kansas, to buy horses for the operation.[28]

All of this riding under very unfavorable conditions caused a boil to form on Horace's buttock, which forced him to stay out of the saddle and treat the eruption with the self-prescribed combination of linseed poultice and Seidlitz powder. This treatment apparently succeeded, for he was back in the saddle after only two days.[29]

Plunkett was back in Cheyenne in early July for the Wyoming Stock Growers Association meeting, where a hot topic on the agenda was the cost of rail freight for cattle shipments. Plunkett took a prominent stand for united action, and the association resolved to try to obtain concessions from the railroads, sending a delegation to Chicago to negotiate the matter. Plunkett was not a member of this delegation but was appointed to a subcommittee to assist this effort. Although this first effort was unsuccessful, Plunkett was happy that he had been asked to work with these "good businessmen," who were also important territorial leaders.[30]

While in Cheyenne in July, Plunkett displayed his skill at chess when he played a lieutenant in the U.S. Army who also had a reputation in the game. To Horace's disgust, the lieutenant "tricked"

[28] Boughton, Plunkett, and Peters each contributed $5,000 to the Tebbetts partnership, which commenced operation on 16 Aug. 1883. Ibid., 15 and 16 June1883, 3 July 1883, and 28 Aug. 1884.

[29] Ibid., 16–18 June 1883. Seidlitz powder, named for the Seidlitz Saline Springs of Bohemia, was a popular laxative.

[30] Ibid., 2, 22, and 23 July 1883. The association committee consisted of Albert T. Babbitt of the Standard Cattle Co, Alexander H. Swan of the Swan Land and Cattle Co., and John N. Simpson of the Continental Cattle Co. The subcommittee consisted of Plunkett, Richard Frewen of the Dakota Stock and Grazing Co., and William C. Irvine of the Ogalalla Cattle Co. *The Cheyenne Daily Leader*, 3 July 1883.

him and won the first game, but Plunkett took the next three "off hand." A week later, he played ecarté for a buggy and harness and won the right to buy them for $80.[31]

Also in July, Plunkett added still another investment to his diverse portfolio. He bought £2,000 of stock in the Brush–Swan Electric Light Company of Cheyenne, which was bringing electric lighting to the Wyoming capital. The company name came from that of the Brush Electric Company of Cleveland and the Swan Incandescent Light Company of New York, both companies being shareholders in the Brush–Swan Company. The Brush Company licensed the generating technology to the Brush–Swan Company, and the Swan Company sold the electric light bulbs used in the system. In this company Plunkett joined a group of prominent Cheyenne men, led by Francis E. Warren.[32]

Early in August, Plunkett became involved in the first of a succession of conflicts with the Frewen brothers when Moreton Frewen repudiated his contract to buy the EK cattle from Plunkett & Roche. Frewen denied the contract obligation by pointing out how foolish it was, saying, "Not one man in 10,000 would act so quixotically." He offered to compromise by taking only 600 head of cattle from Plunkett. Plunkett then submitted the matter to arbitration by their neighbors, Peters and Alston, with Peters taking the Frewen side and Alston representing Plunkett & Roche.[33]

While the dispute with Frewen continued to simmer, Plunkett joined the fall roundup and even learned to rope cattle, catching a total of eleven in his first effort. To add to work and frustration,

---

[31] Plunkett diaries, 18 and 24 July 1883.

[32] The capital of $100,000 was subscribed $80,000 in cash, and the remaining $20,000 was given to the Brush Co. and to the Swan Co. Plunkett's purchase may imply an interest of about 10%, and he gave Alston the option to buy one-third of his stock at his cost plus 12% interest. At the 1884 summer meeting, the company issued $50,000 worth of bonds to its shareholders to meet ongoing capital requirements, and Plunkett took up $2,000 of these bonds. In the spring of 1887, Plunkett held a total of $2,700 of the bonds, and Alston owned $1,300 of them. Francis E. Warren collection, American Heritage Center, University of Wyoming; and Plunkett diaries, 18, 24, and 25 July 1883; 7 May 1884; and 21 June 1884. Although Plunkett soon was discouraged with the Brush–Swan investment, because he thought electric lighting was about to be superseded by gas, the company began to make money, and it paid dividends after 1890. By time the company was sold to the Cheyenne Light, Fuel & Power Company in 1900, dividends of 95% had been paid on the stock, and it appears that the shareholders also received a premium over their original investment

[33] Plunkett diaries, 2 and 20–22 Aug. 1883.

Lord Mayo appeared again, much delayed from his expected arrival because he had bought a new wagon and did not oil the wheels, so the horses played out on the road. Plunkett groused, "Such are globe trotters without niggers to wait on them." Mayo brought along an unnamed guest "of the shop-keeping order," and this "parasite" lost his way in the dark, forcing Plunkett to find bedding for him and six or seven cowboys in the middle of the night.[34]

After the fall roundup, Plunkett instructed the foremen regarding the winter's work, sent the beef cattle herd south to the railroad at Rock Creek, and by September 13 he was ready to leave the ranch, having been there less than four months. The trip south to Cheyenne with young Geoffroy Millais was not as difficult as the journey in May, although the driver got drunk. At the Cheyenne Club, the British members hosted 32 at dinner, and Plunkett was pleased with the evening, saying everyone was drunk, but all were "cordial," no one "beastly drunk." He pronounced the menu "altogether beyond" what he had thought the club capable of.[35]

After this pleasant interlude, the reality of life in the West was waiting for Horace four days later when he traveled west on the railroad to Rock Creek, where he met Windsor and Donahue with the cattle to be shipped east. There was no bed for him, and Plunkett slept on the floor of a railroad construction car; however, his persistent diarrhea sent him back to Cheyenne, where a doctor treated him with opium. Then it was back to Rock Creek to load cattle and to send them off with Donahue and Edmund Roche on board the freight train. This time, Plunkett did not ride in the cattle cars but followed in the passenger train. The Plunkett & Roche cattle sold in Council Bluffs, Iowa, at 4.15¢ per pound, a total of nearly $6,000 less than the Frewen contract would have yielded.[36]

After selling the cattle, Plunkett returned to Cheyenne and met with Tom Sturgis, who wanted him to sell Union Cattle Company stock in England, promising a 5 percent commission to sell $500,000 of the stock. As an additional inducement, Sturgis also offered Horace a position in management, at a salary of $3,000–

[34] Ibid., 28 Aug.–5 Sept. 1883. Plunkett's use of the word "nigger" was in the sense of any dark-skinned person. Queen Victoria scolded Lord Salisbury for calling her Indian attendants "nigger." Kuhn, *Henry and Mary Ponsonby*, 223.

[35] Plunkett diaries, 13–26 and 30 Sept. 1883 and 1 Oct. 1883.

[36] Ibid., 10 Oct. 1883.

$3,500 per year, if he was able to sell at least $250,000 worth of stock. In November, Plunkett began offering the stock to British prospects.[37]

Plunkett suffered a severe personal loss on Christmas Day, when his brother Randal died. With Randal's death the succession to the barony devolved on his brother John, who had a drinking problem. Horace now faced the need to spend more time caring for his father, who gave him a full power of attorney to act for him. He immediately made another strong effort to convince Cyril Coleridge to come to America but to no avail.[38]

As the year ended, Horace faced another cash flow problem, this time in the Windsor, Coble & Plunkett partnership, where cash flow in the second year of operation was even worse than its first year had been. Net income of $38,000 was down from the previous year and again consisted entirely of unrealized gains, because the only cattle sale was one steer for $39.50. All of the operating expenses had to be contributed by the partners. At the end of the year, Plunkett folded all of this partnership's assets into his new Frontier Company.[39]

At the close of 1883, Plunkett's efforts to help Walter Bulwer's son Henry resulted in yet another investment, in an entirely different business, a company in Brooklyn making decorative tiles. This was the period when the decorative tile business, imitating success in Europe, was rapidly spreading across the United States. A few of these companies were quite profitable, but only a handful lasted more than a decade or two, and the one Plunkett invested in was not one of the lucky few.[40]

---

[37] The Union Cattle Co. was organized at the beginning of 1883 to take over the assets of the Sturgis brothers, William C. Lane, and Gordon B. Goodell. Ibid., 14, 21, 22, and 25 Oct. 1883; 7 Nov. 1883.

[38] Plunkett offered to pay Coleridge £1,000 per year, but Coleridge had troubles of his own concerning the financial problems of his five sisters. Plunkett continued to hunt with Coleridge and obtained the latter's help in buying horses. In June 1898, he stopped in Exeter to see Coleridge, then a chief constable: "The best of brothers, as confirmed an old bachelor as I." Ibid., 13 Nov. 1883, 26 and 29 Dec. 1883, 11 Jan. 1884, 23 March 1884, and 20 June 1898.

[39] The Windsor, Coble & Plunkett partnership was liquidated for Frontier Cattle Co. stock valued at $185,200, yielding a book profit of $45,600 to the partnership. Plunkett's share of the liquidation was nearly $67,200 of stock. The partners received stock for their current accounts as well as their capital balances, which accounts for the fact that Plunkett's share was not one-third of the total.

[40] Riley, *Tile Art*, 108.

The International Tile Company was formed in February 1883 by Henry A. Bulwer; John W. Ivery, a civil engineer who had managed a brick factory in England; and Pliny Norcross, a lawyer from Wisconsin. The corporation did not immediately begin operation, but tile manufacturing commenced in a partnership under the management of Ivery. When Plunkett learned of the operation, he arranged for Denis Lawless (a distant cousin) to go to Brooklyn to inspect the business. Lawless returned with the news that young Bulwer was not contributing to the success of the business and should stay at home, but by this time Plunkett, himself, was inclined to invest, although he hoped to limit his liability.[41]

The day after Plunkett landed in New York in the spring of 1884, he went to visit the tile company. A workforce of fifty employees made decorative tiles as well as tiles for floors, walls, and tables. Decorative tiles were made by a technique that minimized distortion of the design by the glazing process, leaving a clear image. Plunkett met John W. Ivery, the manager, and Charles W. R. Wynne, the director. Ivery was then about thirty-one, an English-born engineer, and Plunkett's first impression of him was "sanguine, conceited, energetic," while he thought Wynne was "fairly shrewd, hardworking, unmethodical." A future cause of trouble was the fact that the two managers did not get along well with each other, a situation Plunkett either failed to detect or ignored. Unfortunately, the company was also in serious financial trouble.

Plunkett was tentatively optimistic that the two managers could turn the operation around, despite its near insolvency, because he thought the demand for the tiles was "unlimited." Although he said Lawless was "thoroughly bitten" by the project, Plunkett, if not "bitten," was at least attracted enough to become deeply involved financially in this unfortunate venture.[42]

In 1882 and 1883 Plunkett added more complexity to his business affairs in America, without any recognizable focal point. He successfully raised a good deal of money, but each new scheme required him to raise even more, and profits continued to be elusive.

---

[41] Plunkett diaries, 31 Dec. 1883, 18 Jan. 1884, and 22 March 1884. John William Ivery, who was born in December 1852, managed a brick company in Edgbaston, Warwickshire, before coming to New York in August 1882. Henry Bulwer and his father came to Brooklyn in February 1883.

[42] Ibid., 31 Dec. 1883 and 18 and 22 Jan. 1884; and *Brooklyn Eagle*, 14 Nov. 1886.

5

Among the Wyoming Leaders

When Plunkett was ready to return to the ranch in the spring of 1884, he learned that still another young aristocrat wanted to become a rancher in Wyoming. Windham Wyndham-Quin, heir to the earldom of Dunraven, wrote to Plunkett asking him to take his brother Charles to America. After meeting Charles, who was then twenty, Plunkett wrote to the brother, warning that he would take no responsibility for Charles and would tolerate no trouble if the young man proved not to be the "right sort." Charles went to Wyoming to join the other British ranchers.[1]

Plunkett continued to invest in diverse ventures, and in May, he and Gilchrist launched their first speculation in railroad land. The Union Pacific, which was granted alternate sections in a checkerboard on either side of its line, was now selling that land on easy payment terms. The partners apparently were attracted to this opportunity by the fact the Swan Land & Cattle Company had purchased over 500,000 acres of these lands. Gilchrist bought the first 40,000 acres for the partners in an area adjoining a small ranch owned two-thirds by them, but the chief reason for the purchase was that the land also was adjacent to the range of the Swan Company. Horace called the parcel the "Naboth's Vineyard to the great Swan Co," and he hoped that company would see the threat to its range and buy the contracts at a good profit to the partners.[2]

[1] Plunkett diaries, 12 and 14 April 1884.

[2] Ibid., 8 May 1884.

Plunkett apparently gave little thought to the risk of offending Alex Swan, who had organized the Swan Company in Scotland and was paid the princely sum of $10,000 to manage the huge ranch. Swan was at the pinnacle of cattle ranchers in Wyoming, and he was a formidable figure in the territory, although he was not a member of the group Plunkett associated with. In banking terms, the men who organized the Wheatland Project were part of the Stock Growers National Bank faction, whereas Swan was a member of the First National Bank faction in Cheyenne. Swan later tried to frustrate Plunkett and Gilchrist's land speculation, but there is no evidence that their initial land speculation damaged Plunkett's relationship with other Wyoming leaders.[3]

Swan did agree to buy contracts for 50,000 acres from Gilchrist at $1.50 per acre, but then he also tried to avoid giving Plunkett and his partners such an easy $25,000 by asking the railroad to void the Plunkett contracts. Fortunately for Plunkett and Gilchrist, Swan's efforts with the railroad failed. In July 1884, the Swan Company purchased another 24,000 acres of railroad land from Plunkett and his partners at the same price as the previous purchase.[4]

This purchase and quick resale only whetted Plunkett's appetite for more railroad land speculation, and soon he and Gilchrist purchased an additional 95,000 acres from the Union Pacific, this time with Boughton as an additional partner. The contracts with the Union Pacific were all at $1 per acre and called for payment over ten years, with interest at 6 percent. The three men then formed the Ione Land & Cattle Company, capitalized at $2 million, to take over their small ranch and to manage 54,000 acres of railroad land.[5]

Boughton managed the Ione Company, which removed him from the Frontier Company without the need for the company to buy his shares. Plunkett was only a minor shareholder in the Ione Company, and in practice he visited the ranch only when Boughton was ill. For

---

[3] The First National Bank in Cheyenne was organized in 1870 with Amasa R. Converse as president. After Alex Swan came to Cheyenne, he became a shareholder and director of the bank. The Stock Growers National Bank was organized in 1881 by the Carey brothers, the Sturgis brothers, and Henry Hay. Joseph Carey and his brother initially owned a majority of the stock. Woods, *Sometimes the Books Froze*, 21, 28–29.

[4] Plunkett diaries, 6 and 19 June 1884, 28 July 1884, and 1 Aug. 1884. The chairman of the Swan Co. was irate at Swan's purchase of Plunkett and Gilchrist's lands, calling the deal blackmail. Colin J. Mackenzie's report to the Swan Co.'s board, 4 Nov. 1884, Western Range Cattle Industry Study, Colorado Historical Society.

[5] Plunkett diaries, 12 May 1884.

his part, Boughton welcomed the opportunity to run his own operation on the Laramie Plains, with only minimal direction from Horace. Adopting a grand stance reminiscent of Moreton Frewen, Boughton declared that he would to take no salary from the company for at least four or five years, and probably not even then, as he expected to realize his remuneration from the stock he held.[6]

The Ione Company occupied a large block of land east of the Union Pacific railroad line, near the Hutton Station, some twenty-five miles west of Laramie. The Ione ranch did not own many cattle and was really primarily another irrigation project that required large investments to bring water to the land. The shareholders also hoped that they would find valuable deposits of iron ore and plumbago (graphite) on the land, which would substantially enhance the company's value.[7]

A source of annoyance for Boughton was the face that the railroad had permitted others to construct improvements on some of the land transferred to the Ione Company. Gilchrist and Plunkett agreed with the railroad that they would sell these parcels to the owners of the improvements, at prices generally consistent with the railroad's policy on such sales. Nevertheless, dealing with these occupiers was an awkward and time-consuming problem. A fairly extended dispute was with John H. Douglas-Willan of the Douglas-Willan, Sartoris Company over the improvements that company had made on the railroad land.[8]

---

[6] At the time the Ione Land & Cattle Co. was organized, Boughton bought two-thirds of the stock in the company, paying some $35,000 in cash and giving his note for an additional $20,000. As he expected the total working capital requirements to be £22,000–£25,000, he at once set about raising more money from his family and friends. One who ultimately became a shareholder of the company and also worked on the ranch was young Fitzgerald C. "Fritz" Peploc, the son of Major Peploc; Boughton made a similar deal with A. H. Pryce, son of R. D. Pryce of Welshpool, Wales. In the summer of 1886, Boughton's brother-in-law, Sir Offley "Johnnie" Wakeman, from Shrewsbury, England, also became a shareholder of the Ione Co., as did John Whitaker, Jr. Whitaker later purchased the ranch when Boughton sold out. E. S. R. Boughton to Major Peploc, 27 Oct. 1884, Ione Company letterpress, on microfilm in the Western Range Cattle Industry Study at the Colorado Historical Society (hereafter cited as Ione letterpress).

[7] The cattle tally in the fall of 1886 was only 747 head. E. S. R. Boughton to Major Peploc, 27 Oct. 1884; E. S. R. Boughton to John J. Linck & Co., St. Louis, 21 Oct. 1884; and Samuel Corson to A. H. Pryce, 26 Sept. 1886. Boughton said that Iron Mountain was "a solid mountain of iron ore." E. S. R. Boughton to R. D. Pryce, 18 Oct. 1884. Ione letterpress.

[8] E. S. R. Boughton to Lionel Sartoris, 18 Oct. 1884, Ione letterpress. The claim by J. H. Douglas-Willan was pursuant to some sort of lease with the railroad and was not an insignificant matter, for he asked to purchase three and a half sections. J. H. Douglas-Willan to Andrew Gilchrist, Horace Plunkett and E. S. Rouse Boughton, 15 Nov. 1884, Ione letterpress.

Having resolved his relationship with Boughton by installing him in the Ione ranch, Plunkett turned his attention to the Powder River ranch, which was not in good order. As usual, when he arrived, there was nothing to eat, "everything" had been stolen, and in general the place, which was presided over by the sulky Jennings couple, was filthy. Moreover, local management on the range was also a problem, because the first foreman, Jack Donahue, who had seemed so promising three years earlier, was now abusing his horses, had no respect for others, and was very intractable. Although Plunkett believed that Donahue was honest ("as far as I know"), Donahue was also a poor judge of cowboys and had hired one of the worst thieves in the county, which Plunkett feared would cause trouble with other cattlemen.[9]

Out on the range, there was more trouble. On the way to the Powder River roundup early in June, Plunkett found that high water in the river prevented the roundup crews from crossing. About 100 cowboys were camped by the river, amusing themselves by racing their horses and playing cards. From across the mountain came a report that the Shoshonis, some 500 strong, had left the reservation and were in the Nowood country. However, when the Indians passed the roundup, where beef cattle were being killed, they did not ask for their "usual fifth" share of the meat. An Arapaho came to Plunkett's ranch to beg for food, and they had a long talk in sign language so that Plunkett could learn the man's account of conditions with the Indians. The Arapaho claimed members of his tribe were killing buffalo across the mountains, whereas the Shoshonis were killing cattle. He mentioned that two of Jim Winn's cattle had been killed, and he described the brand for Plunkett.[10]

Plunkett rode over the mountains, which was still snow-covered at 8,000 feet, to Jim Winn's home ranch (the Big Horn Cattle Company, using the Bar X Bar brand), and there he found the ranch house Frank Ainsworth was building by the waterfall on Canyon Creek. In the spring of 1883, foreman Frank Bull had married a lady's maid who came to Frewen's ranch the previous Octo-

[9] Plunkett diaries, 9 May 1884.

[10] Ibid., 3 and 5 June 1884. The Shoshonis were forced to accept the Northern Arapaho tribe on the Wind River Reservation, making for a very uneasy relationship between the former enemies. Consequently, the Arapaho's story of Shoshoni complicity in cattle theft may not have been reliable.

ber and stayed to spend the winter; she was the only woman on the ranch. In October 1883, Bull went to work for Jim Winn's Bar X Bar outfit and brought his bride across the mountain to the unfinished house, where the couple spent that winter in a nearby unchinked cabin that admitted both the cold and rats from outside. In March 1884, the Bulls moved to the one finished room in the big house (the rest of the house still lacked a complete roof).[11]

At the time of Plunkett's visit, Mrs. Bull was about to go to Buffalo to be delivered of her first child, a journey that normally was formidable and soon became downright harrowing. On their way across the mountains, the Bulls came upon the body of a man who had been dead some days. Afraid for her life, Martha Bull would not let Frank examine the body, but on the way to Buffalo they met Frewen's foreman, Fred G. S. Hesse, and reported the death.[12]

Hesse reported the Bulls' account to Frewen, who sent a man to investigate: the dead man had been shot twice and then dragged into a gulch. Frewen and the coroner from Buffalo went to the scene, where an inquest was held and then the corpse was covered with stones and wood because the terrain was too rocky to dig a grave. Frewen surmised from a few Indian beads in the man's pocket that the victim had stolen an Indian horse and been shot for the offense. Plunkett offered four possibilities: (1) the man was a stock detective, (2) he was a horse thief killed by an accomplice, (3) he was killed by Indians, or (4) he was a deserter from the Army. The inquest concluded there was no evidence to reach a conclusion.[13]

Plunkett developed a very bad sore throat, and on his way to Cheyenne he stopped at Fort Fetterman to see a doctor practicing

---

[11] Ibid., 12 June 1884; and Frison, *First White Woman*, 7–21. Mrs. Bull, the former Martha James, came to the United States in April 1883 with William C. Armstrong, who brought his bride to honeymoon and visit friends in Cheyenne before going up to Frewen's ranch in October. Martha James then stayed behind on the Powder River. Mrs. Bull's recollections were recorded by Frison in 1935, more than 50 years after the fact, and are flawed in some details, but they provide a valuable window on her experiences at the Bar X Bar Ranch.

[12] Gwendolyn Bull was born 16 July 1884. Frison, *First White Woman*, 18, 21. Mrs. Bull claimed that her daughter was born in the Big Horn Basin, the first child to be born there, but Moreton Frewen's journal entry of June 22 said that she was on her way to Buffalo to be delivered when she and her husband saw the dead body. Frewen collection.

[13] Moreton Frewen's 22 June 1884 entry, Frewen collection; and Plunkett diaries, 20 June 1884. Mrs. Bull said that the coroner's jury thought the dead man was the roundup cook, who had been missing for ten days, but the contemporary accounts do not mention this possibility.

there. The doctor, who was the roughest looking practitioner Plunkett had ever seen, was treating a hospital full of "broken & bullet-pierced cowboys." As to Plunkett's throat, the doctor said it was "a Hell of an old throat," and when Horace asked for more details, the doctor added that it was "diphtheria," but that he would "scatter" it. He gave Horace some powders, which the latter thought was mercury, and for whatever reason, the throat improved, and three days later was entirely cured.[14]

At the Cheyenne Club at the beginning of July, Plunkett dined with John Lubbock, the son of a baronet, and this time he was uncertain whether to wear "evening" or "Sunday" clothes for the occasion. He mused that only a year earlier, a flannel shirt would have been de rigueur. On the Fourth of July, which he regarded as a day of humiliation for the British, Plunkett summarized the celebration as consisting of processions, speeches, races, religious exercises, fireworks, and whiskey. While observing all of this, he conceived yet another idea for a possible Wyoming investment—that of a stock exchange. However, no more was mentioned of this notion.[15]

From Cheyenne, Horace went to the Wheatland Project to inspect the tunnel being bored to bring irrigation water from the Laramie River to Bluegrass Creek. A difficult engineering problem, the tunnel was to convey water from the Laramie River, through the divide between the two streams, from whence it was conveyed to the company's lands via the Sybille. When Plunkett saw the work in July, the tunnel reached 1,000 feet into the mountain, but another 2,100 feet needed to be bored to reach the other end. The next year, the workers began boring from both ends, using a water-powered air compressor, and progress was considerably improved.[16]

Plunkett finally organized the Frontier Cattle Company in the middle of July 1884, with authorized capital of $1.5 million. He was president of the company, and Gilchrist was vice president and

---

[14] Plunkett diaries, 23 and 26 June 1884.

[15] Ibid., 1 and 4 July 1884. One may assume that the dinner companion was John Lubbock, because Plunkett said he was dining with the son of Sir John Lubbock, and the baronet's eldest son was named John.

[16] The contract for the tunnel called for a bore of 3,100 feet, at a cost of $15 per foot, and the total cost of the tunnel and associated dam was over $230,000. Ibid., 8 July 1884 and 23 and 24 Aug. 1885.

manager of range operations. The active shareholders, in addition to Plunkett, included Edmund and Alexis Roche, Andrew Gilchrist, Henry J. Windsor, J. E. "Dude" Booth, Henry P. Maxwell, and John Chaplin, who was Plunkett's agent in Cheyenne. Others, such as William Blacker and Boughton (a "large" shareholder), were intended to be purely passive investors. William Douglas Watson-Smythe, a schoolmate from Eton, purchased shares and wanted to take part in the business, and although Plunkett accepted his money, he grumbled that Watson-Smythe would probably be a "troublesome" partner.[17]

Initially, the Frontier Company included only the properties of the Plunkett partnerships. These were Plunkett & Roche (the EK and NH ranches); the Windsor, Coble & Plunkett partnership; and Andrew Gilchrist's ranch, which was down on Crow Creek, near Cheyenne. In August, the Frontier Company took over the railroad lands remaining after the transfer to the Ione Company, and it also purchased James Allen's ranch on the South Sybille.[18]

While still in Cheyenne after the organization of the Frontier Company, Plunkett dealt with details of some of his other diverse business interests. He attended a meeting of the Brush–Swan Electric Light Company, where the shareholders agreed to contribute additional money, his share being $2,000. Then he helped renegotiate the compensation of the contractor for the Wyoming Development Company's Wheatland Project. Finally, Plunkett joined in the Wyoming Stock Growers Association's new effort to reduce freight rates for cattle. The first delegation, headed by Alex Swan, came away empty handed. This time, Plunkett was a full member of the delegation that went to Denver to meet with the railroad.[19]

The new group consisted of association secretary Sturgis, associ-

[17] Cattle were contributed to the Frontier Co. at $35 per head (with the calves thrown in at no additional value), except for purebred Hereford stock, which were taken in at $75 per head for mature stock and $60 for the calves. A further $2 per head was paid for range rights and property improvements, and horses and equipment were taken at an appraised value. The fee land in Gilchrist's Crow Creek ranch was valued at $20 per acre for the hay land, and $2.50 per acre for the railroad land. Ibid., 13 and 17 July 1884 and 26 Feb. 1886.

[18] The railroad lands were purchased at $1.50 per acre, and the company paid $40,000 for the Allen ranch. Allen was born in England in January 1847. Ibid., 8 and 9 July 1884; and Burns, Gillespie, and Richardson, *Wyoming Pioneer Ranches*, 366.

[19] Plunkett diaries, 21 and 23 July 1884.

ation president Carey, John N. Simpson, and Plunkett. Of the four, only Simpson was a member of the first delegation, and the fact Plunkett was asked to join the new effort is a clear indication of his stature in Wyoming. Plunkett said that he took an active part in the several meetings on that occasion. He told the railroad representatives that he represented the large area north of the Union Pacific and south of the Northern Pacific, "& practically controlled the situation." Whether or not the railroad was impressed by concerted action on the part of so many shippers is unknown, but in any case the association soon reported that the Union Pacific was granting a 5 percent discount from 1883 rates for shipments between Ogden, Utah, and North Platte, Nebraska.[20]

With the consolidation of several of his partnerships into the Frontier Company, Plunkett's management responsibilities had grown, and when he got back to the ranch, he had to face up to the shortcomings of his organization. Donahue continued to be a problem, for although he was without peer working the wagons and was honest, he was also careless, was a poor judge of men, and neglected anything not involving cow punching. When Plunkett refused to make him the overall foreman, Donahue wouldn't accept a lesser role, and he had to go, although Plunkett hoped that Donahue would work a while to give du Fran a chance to settle into the job. Unfortunately, the next day Plunkett found Donahue fomenting trouble with the other hands and fired him on the spot. On the occasion of Donahue's departure, after four years in the job, Plunkett summarized his opinion of the man as follows: "[A] desperado by nature and by education . . . terribly profane & blasphemous at times—[but] was generally amusing. He thoroughly understood the expressions of the western language & some of his sayings will long be remembered by Plunkett Roche & Co."

Plunkett didn't want to replace Donahue with a man working in the area, but no other alternatives were available. In the end he

[20] The association notice of the agreement with the Union Pacific was dated Aug. 1, 1884. The agreement also gave the shippers flexibility in choosing to sell cattle among Council Bluffs, Iowa; Omaha, Neb.; and Chicago, Ill. The railroad emphasized that its concessions were made "voluntarily." *Cheyenne Daily Sun*, 3 Aug. 1884; and Plunkett diaries, 23 July 1884.

gave the job to Phil du Fran, even though he had contemplated firing du Fran just days before he made the decision to terminate Donahue.[21]

Over the mountains at the Winn ranch in August, Horace found an alarming situation where three horses had suddenly developed badly swollen heads on the same day. Plenty Bear, the noted Arapaho head man, lanced the swellings, declaring they were snake bites. Plunkett was skeptical, considering it unlikely three horses would be bitten in a single day. However, he wrote, "the 'untutored' was correct," for the horses recovered as the result of Plenty Bear's treatment.[22]

Always concerned about his health, Plunkett weighed himself in August at 132 pounds with his clothes on (which he estimated added eight pounds). This was his lowest weight in years and was down from the middle of June, when his naked weight was 130 pounds—not very robust for a man five feet ten inches tall. Still, he had considerable physical stamina, and to while away the time on a rainy day, Horace bet that he could walk a measured mile in twelve minutes. Watson gave him odds of nine to seven on the bet, and Plunkett won $45 with a time of ten minutes and ten seconds.[23]

The only town of any size near the Powder River ranches was Buffalo, the seat of Johnson County. In August, Plunkett set down his description of the town, which he had first seen three years before in 1881—it was now a thriving place of 600: "The town is beautifully situated on both banks of a beautiful mountain stream (Clear Creek). The Occidental hotel was a regular gambling Hell. Monte, pharo, keno & other iniquitous games occupy the large hall of the building while whiskey & music make night hideous for

[21] Plunkett hired du Fran at $125 per month, effective when Donahue left, but he told du Fran he would only be paid $100 a month in case the company decided to place another foreman over him. Plunkett diaries, 1 and 3–7 Aug. 1884. Philip du Fran was born in Iowa in May 1854. Plunkett also had management problems in the Tebbetts & Co. partnership, where he attributed the financial failure to Tebbetts's incompetence.

[22] Ibid., 13 Aug. 1884. Plenty Bear had a long history of friendship with the government. He had favored peace after the Custer battle in 1876, and the following year he visited Washington, D.C., with Chief Black Coal. He was said to be an imposing figure, over six feet in height, weighing 250 pounds, with the countenance of a Roman senator. Walker, *Stories of Early Days in Wyoming*, 196.

[23] Plunkett diaries, 16 June 1884 and 17 Aug. 1884.

those who do not drink, gamble or swear. They will have hard times for a while before they settle down to prosperity."[24]

Serious troubles soon erupted again with Alexis Roche, who was no longer a part of management but still felt his opinion should be honored. Alexis had gotten into an argument with an employee called "Old Pete" and wanted Plunkett to fire Pete, but Plunkett declined to do so. In his diary, Plunkett explained that Pete was wrong, but that his incivility stemmed from Alexis's ill treatment of the man. He said that Alexis disregarded the feelings of everyone except himself, but he was very touchy when his own feelings were offended. After the initial confrontation, the matter festered with Alexis for two weeks, and he accused Horace of trying to drive him out of the company, which gave rise to another confrontation, after which Alexis retracted his accusation.[25]

Responding to rumors that the Desert Land Act might be repealed, Plunkett accelerated his efforts to file on land so as to control the cattle range. The big ranches hired "reliable" men (usually ranch hands), paying them $25 at the time of filing and $25 when final proof was made. Once the patent was issued, the owner was expected to transfer the land to the company. In order to prove up on a desert land entry, the land had to be irrigated, and Plunkett personally went out to locate ditches in the area where the filings were to be made, both in the Powder River region and on the Laramie Plains. The Blue Creek Ditch received water in mid-September 1884, and the Boughton and Windsor ditches opened later in the same month, all from the North Fork and Middle Fork of the Powder River.[26]

---

[24] It is interesting that Plunkett did not see Buffalo before 1881, because the town was founded in 1879, the year he arrived in the Powder River Basin. However, Frewen's store and post office on the Powder River dated from May of 1879 and undoubtedly sold such goods and services that could be found in the remote region, making the trip to Buffalo unnecessary. Buffalo's post office was opened in October 1879, and the Occidental Hotel building was completed during the fall of 1880. Lott, "Old Occidental," 27; and Plunkett diaries, 20 Aug., 1884.

[25] Plunkett diaries, 31 Aug. 1884 and 15 Sept. 1884.

[26] The description of the filing system then in use by both the Frontier Co. and the Ione Co. is from Corson to E. S. R. Boughton, 15 Feb. 1885, Ione letterpress; and Plunkett diaries, 18 and 19 Sept. 1884. The Blue Creek ditch to irrigate 161.4 acres was from the North Fork of the Powder River, as was the Boughton ditch, which was to irrigate 855 acres, with a priority of Sept. 25, 1884. The Windsor ditch had the same priority date, and it was to irrigate 86 acres from Beaver Creek, a tributary of the Middle Fork of the Powder River. Despite Plunkett's record regarding water in the Blue Creek Ditch, the official priority for it is 15 July 1886 *Tabulation of Adjudicated Water Rights: Water Division Number Two*, 136, 141, 142.

Down on the Ione Company lands, Boughton also hired men to file on the government land interspersed with the Ione Company railroad sections, and by the fall of 1884 he had already paid for filings on six and a half sections. With his typical arrogance, he brushed aside any problems with this system of acquiring land, saying that if the title proved defective he would still be "in command" of the situation because he had the only source of irrigation water. To further extend his control over the government sections, Boughton also expected to fence the government sections, saying he could do that without getting off the railroad land.[27]

The irrigation works on Ione Company's property were much more extensive than those Plunkett was building in the Powder River drainage in the north. The Boughton ditch from the Laramie River was about nine feet wide, and the main ditch was about four miles long, with laterals extending another eight to ten miles, to irrigate nearly 7,900 acres. Opened in September 1884, this ditch was fed from a dam across the Laramie River. The only Wyoming ditch named for Plunkett was opened from the Middle Chugwater Creek a year later.[28]

As he contemplated the heavy expenditures that would be needed to irrigate the lands he wanted to file on along the North Fork of the Powder River, Plunkett began to worry whether the land would be worth as much as the cost of acquiring it. Moreover, a negative factor he had not expected was the possibility that oth-

[27] E. S. R. Boughton to R. D. Pryce, 18 Oct. 1884, Ione letterpress. It was illegal to fence government land, and Boughton's idea of enclosing the government sections in the checkerboard by fencing the private sections ultimately failed, although it would be some years before the final answer came down from the U.S. Supreme Court. His concept of controlling land by controlling irrigation water was also faulty, because the water was appropriated to the land, not to the person who built the ditches.

[28] The Boughton ditch from the Laramie River had a priority of 1 Sept. 1884 to irrigate 6,844 acres, and a further 1,038 acres were added on 5 Oct. 1884. Ione letterpress. The Plunkett ditch from the Middle Chugwater Creek had a priority of 14 Oct. 1885 to irrigate 400 acres. *Tabulation of Adjudicated Water Rights: Water Division Number One*, 26, 39. The Ione letterpress mentions two other ditches, the Pryce ditch and the Plumbago ditch, but I have found no other reference to them. Boughton wrote to Lyulph Ogilvy, the younger brother of Lord Airlie, for help in breaking 640 acres of prairie sod to plant alfalfa. It is unknown whether Boughton sought help from Ogilvy because of the title or whether Ogilvy actually had some practical experience of that sort on his farm down in Colorado. E. S. R. Boughton to Lord Ogilvy, 9 Nov. 1884; and E. S. R. Boughton to R. D. Price, 18 Oct. 1884, Ione letterpress. An interesting family connection of Lyulph Ogilvy was his sister, Lady Henrietta Hozier, who was the mother of Clementine, the future wife of Winston Churchill.

ers would file on land that could be irrigated from the ditches he was building. Early in 1885, he learned that in the NH Ranch pasture, others had filed on the land he expected to irrigate, and the land with the ditches now belonged to these earlier filers.[29]

In September 1884, the death of Gillie Leigh cast a long shadow over the young men ranching on the Powder River. Gilbert Henry Chandos Leigh, who had gone to Cambridge with Moreton Frewen, was heir to the barony of Leigh and was a member of Parliament at the time of his death. Leigh first visited Wyoming Territory when hunting with the Frewen brothers in the fall of 1878. This was the trip on which the Frewens later crossed over to the Powder River Basin to select the location of their ranch.[30]

Leigh returned more than once to visit the area, and in the fall of 1884, he was with a hunting party in the Big Horn Mountains when he left camp alone, apparently to look for Bighorn Sheep. Because he liked to hunt alone, his companions were not immediately concerned when he did not return to camp. Finally, a search for him was launched, but the place where he had fallen off the canyon wall was not discovered for an entire week after his death.

Plunkett was fond of the good-natured, genial Gillie, whom he called "the rascal," and he was shocked by Leigh's death. Plunkett was on his way to Cheyenne when he learned of Gillie's death, and in Cheyenne he met Leigh's younger brother, Dudley; the two then went back up the line to Rock Creek to meet the remains. Both the Buffalo undertaker and one from Cheyenne were at Rock Creek, and after Plunkett vainly tried to have the remains treated with more respect, the body was loaded into the baggage car for the sad journey home.[31]

In mid-October, the annual meetings of both the Frontier Company and the Wyoming Development Company were held in

---

[29] Plunkett diaries, 15 Jan. 1885. Plunkett referred to the "HN," obviously a transposition of "NH."

[30] Gilbert Leigh sat in the House of Commons from South Warwickshire, and his father sat in the House of Lords.

[31] Plunkett recorded the news of Leigh's death on 23 Sept. 1884. Plunkett must have raised a considerable fuss with the Union Pacific, because the railroad agent used the incident as one of the reasons for denying Boughton a pass to travel on the railroad. E. S. R. Boughton to Mr. Kimball (Omaha), 9 Nov. 1884, Ione letterpress; and Plunkett diaries, 23 Sept.1884 and 5 and 6 Oct. 1884.

Cheyenne. Plunkett said all the shareholders were present for the Frontier Company meeting, and Chaplin was elected secretary of the company, which provides further evidence of Plunkett's confidence in him. At the Wyoming Development Company, Carey was elected president; Plunkett was vice president; Sturgis, who had succeeded Warren as a shareholder, was secretary; Gilchrist was general manager; and Post was treasurer. The company thus continued the façade of Carey as the leader, while Horace Plunkett and Gilchrist ran the operation in the background.[32]

Although Carey was the largest shareholder and perhaps the richest resident of the territory (at least after Alex Swan's insolvency), Plunkett never hesitated to lecture the judge. There was a genuine friendship between the two men, and Plunkett even seemed to approve of Carey's custom of enforcing his teetotaler views on his dinner guests. For his part, Carey always acknowledged the company's indebtedness to Plunkett's efforts on its behalf.[33]

In October, Plunkett was in charge of making the arrangements for the testimonial dinner the Wyoming Stock Growers Association held for Tom Sturgis, honoring him for his service as secretary of the association. For the powerful association to trust an Irishman with this task was an extraordinary mark of favor for Plunkett. He said, "I had the pleasure of pushing this well deserved tribute through & completing all the arrangements without the pain of having to appear in public." Both Plunkett and Gilchrist were on the stage for the occasion, and Sturgis singled them out for specific mention in his remarks accepting the gifts.[34]

On the Ione ranch, Boughton emulated the other ranchers' tradition of leaving the place in the hands of hired employees and

---

[32] Plunkett diaries, 13 Oct. 1884.

[33] Ibid., 16 Oct. 1884.

[34] The dinner, held at the Opera House, was "generally" full formal dress. Ibid, 15 and 16 Oct. 1884. Both Carey and Sturgis referred to Plunkett by name, although the *Cheyenne Sun* spelled his name "A. C. Plunkett." The *Cheyenne Leader* omitted the names of Plunkett and Frederick de Billier from the *Sun* list of those on the stage, which would lend credence to Plunkett's statement that he had not appeared in public on the occasion. The presentation included a gold-lined silver punch bowl and two silver candelabra from Tiffany's (worth a total of $5,000) and the painting "Courier's Shot at the Wolf" by the Russian artist Ivan Ivanovitch Kowalski (valued at $2,000). Col. Albert T. Babbitt was delegated to select the gifts, but Plunkett paid for them in New York on his way home to Ireland. *Cheyenne Sun,* 16 Oct. 1884; and *Cheyenne Democratic Leader,* 16 Oct. 1884.

departing for home across the Atlantic. He designated young Fritz Peploc as his representative on the ranch for the winter, loftily assuring the young man's father that there would be a "respectable" married couple in the house, plenty of horses to ride, and good antelope shooting. Unfortunately, before the end of the year, the cook left the ranch, apparently because the authorities in Laramie were looking for him "for some stabbing business." A replacement was found but only at a substantial increase in wages. As the winter wore on, young Peploc found the place "rather dull," and Samuel Corson, the company bookkeeper in Cheyenne, sent out some novels for him to read. Corson also bought a small alarm clock for the young man, who apparently was not sufficiently motivated to arise in the morning. One good result of the long "dull" isolation was that Fritz finally got "the hang" of bookkeeping.[35]

In New York before leaving for Ireland, Plunkett learned that Ivery and Wynne were at loggerheads in the tile company, which once more had run out of money. Unable to resolve the management problem, Plunkett nevertheless met with Walter Bulwer in November, and the latter agreed to try to raise additional money in London for the Brooklyn, New York, tile company. Plunkett worked with Bulwer on a prospectus to raise the money, and finally Horace committed a further £2,000 to the project. At the end of the year, $35,660 had been subscribed, and $20,000 was cabled over to Brooklyn.[36]

When Plunkett returned to New York in the spring of 1885, he attended a meeting of the tile company shareholders, who agreed to give old Bulwer full powers to act for the directors. Unfortunately, the Brooklyn management was still in turmoil, and Plunkett and Bulwer had to advance £3,000 to keep the company from failing. Still, the company continued to struggle, and at the end of the year more money had to be raised to avert bankruptcy. Nevertheless, Plunkett still believed in the long-term prospects for the business, and he brought in new investors. One of those new

[35] The cook's wages had to be raised to $50 from $35 per month. E. S. R. Boughton to Major Peploc, 18 Oct. 1884; and Samuel Corson to E. S. R. Boughton, 9 March 1885, Ione letterpress. Corson's full time-employment was with the Union Mercantile Co. in Cheyenne.

[36] Plunkett diaries, 21 October 1884; 14, 25, and 27 Nov. 1884; and 29 Dec. 1884.

investors was Fred Verney, the younger son of a baronet (who turned nasty when the company again began to fail).[37]

Wyoming ranchers had talked of the risk of overcrowding for years, and although range conditions continued to deteriorate each year, 1885 was the first year of concerted action to do something about it. Because nothing could be done to restrict others from entering the range, the big outfits had to reduce their costs, as the poor condition of the herds began to reduce revenues. The years ahead were difficult, both for the owners and the cowhands.

---

[37] On his way west, Plunkett visited the St. Paul, Minn., agent of the tile company and asked a New York manager to come to St. Paul to deal with the poor customer relations. Plunkett was able to raise an additional £8,000, apparently in the form of debentures, of which £6,000 came from Lord Dunsany and £1,000 from old Bulwer. Ibid., 20 April 1885, 20 Oct. 1885, and 7 and 24 Nov. 1885. Frederick William Verney was born in 1846.

6

The Liquidating Manager

Although Plunkett had plenty of personal problems in 1885, that was the year in which he became directly involved in running Frewen's Powder River Cattle Company. When Moreton Frewen incorporated the Big Horn Ranch partnership as the Powder River Company, he got the money to buy out his brother Dick and to go on a buying spree for more cattle and even some sheep. This uncontrolled expansion soon exhausted the cash resources of the company, and when the London board ordered him to cut back his purchases, Moreton was incensed.

Moreover, routine ranch management palled on Moreton, who had so many other things he would rather do. Accordingly, in March 1885, Moreton asked Horace to take over management of the Powder River Company from him at a salary of £1,000 per annum. Of course, Plunkett's range was adjacent to that of the Powder River Company, so he already knew a great deal about that unfortunate company's affairs. Although he could see that it would make sense if range management could be coordinated between the two outfits, he nevertheless refused to manage the company, chiefly because he was worried that his health would not tolerate the added strain.[1]

---

[1] In mid-January, Plunkett attended the annual meeting of the Powder River Cattle Co. in London, where he apparently had been invited to speak on the condition of the American cattle business. Plunkett did agree with Frewen that their two operations should coordinate the wage scale for cowhands and cooperate by sending a portion of their herds to the mountains to reduce overcrowding on the range. Plunkett diaries, 15 Jan. 1885 and 27 March 1885; and Horace Plunkett to Moreton Frewen, 21 May 1885, Frewen collection.

Plunkett's relationship with the Frewen brothers was complicated. In his diary, he was critical of their operation, saying that the brothers had lots of ability but no talent, and he also sneered at their lack of feeling for the peculiar sensitivities of the western people, which made them unpopular. Nevertheless, he also functioned as a sort of father confessor to the Frewen brothers (as he did with many other people).[2]

Plunkett landed in New York on April 19, 1885, and after stopping in Brooklyn for a meeting of the tile company, he took the train across southern Canada to visit the Powder River Company's feeding operation at Superior, Wisconsin. This facility was another brainchild of Moreton Frewen, but it was not a favorable location to feed cattle as there was no cheap source of feed nearby. Nevertheless, Horace felt it could be a good site for shipping cattle eastward through the Great Lakes. Ever mindful of land speculation opportunities, Horace thought that if lake traffic increased, land prices around the town would increase.[3]

After inspecting the Superior facility, Plunkett turned his attention for a time to his own diversified business interests. From Wisconsin he went to Omaha to check on the real estate market, and there he was introduced to the American pastime of baseball. His description of this strange entertainment is characteristic of his style: "Went to see a baseball match & had the game explained. It is much simpler than cricket, it requires more agility & strength & less skill. It only lasts from 1 to 2 hours instead of days & so is better for business men. Players were all professionals. Much excitement at the match."[4]

In Cheyenne, Plunkett had to deal with a management problem in his Frontier Company. As noted earlier, Gilchrist was supposed to manage the range operations of the company, and for this work he was to receive half of Plunkett's salary. Most of the company's range was in the north, but Gilchrist continued to live in Cheyenne, and he tried to delegate his range responsibilities to Windsor. Plunkett refused to authorize this delegation of responsibility, and Gilchrist became petulant, complaining in effect that Plunkett did not have the right of veto over the decision. Gilchrist only

[2] Plunkett diaries, 7 and 24 Nov. 1884.

[3] Ibid., 30 April 1885 and 1 May 1885.

[4] Ibid., 3 May 1885.

capitulated after Horace wrote him a long letter. Gilchrist then said he was the "vanquished foe," and Plunkett noted in his diary, "This is manly."[5]

On examining the books of Frontier Company, Plunkett found the company had lost a good deal of money during the winter. He realized that the company needed a cattle-feeding operation in Nebraska, but he also set about cutting expenses. To improve control and to cut costs, he told the merchants in Buffalo that the ranch hands could no longer charge their personal purchases to the company. As he reduced expenses, Plunkett soon realized the company was overstaffed, for there were simply not enough jobs for all the men he had gathered around him. Dealing with this problem involved more than simply eliminating jobs, because he fretted that the Roches, Maxwell, Booth, Watson, Boothy, and Wyndham-Quin needed his help to get into "some profitable scheme." Unfortunately, Plunkett's career planning for these young men continued to be costly and did not produce positive results for the men, either.[6]

Plunkett's hurried choice of Phil du Fran to replace Donahue as foreman came back to haunt him after only ten months, for work on the range was suffering. At the end of May, he decided that Phil du Fran would have to go, and the two men went to Buffalo to settle accounts and to unravel the winter's accounts for the ranch. To replace du Fran, Plunkett made another snap judgment and chose Roach Chapman, who was 22. Although Plunkett admitted there was a question about Chapman's character, the early signs were all positive, and for a time Chapman earned high ratings in Plunkett's diary.[7]

Plunkett hired two new two employees who did have considerable promise. The first was William "Beau" Watson, who shared a long intimate friendship with Horace. Beau was then 20, the son of Robert Gray Watson of Ballydarton, a celebrated huntsman and

---

[5] Ibid., 1 and 13 Jan. 1885 and 22 July 1885.

[6] Ibid., 5, 6, 10, 12, and 25 May 1885.

[7] Ibid., 30 and 31 May 1885 and 4 and 5 June 1885. At the time of the Johnson County invasion of 1892, Phil du Fran was a detective for the Wyoming Stock Growers Association, and he accompanied the cattlemen's invasion party, although he was not charged in the subsequent prosecution of the invaders. According to a subsequent newspaper account, Chapman's name was U. G. Chapman, perhaps Ulysses Grant Chapman, who was apparently born in Illinois in June 1863. *Big Horn Sentinel*, 15 May 1886.

Plunkett's near neighbor back home in County Meath. Soon after hiring Watson, Plunkett rode out to the Crow Creek ranch, where he found young Watson had plunged too enthusiastically into the unaccustomed range work and was now laid up with a bad chill. Horace took him to Cheyenne and placed him under a doctor's care. Ten days later, Beau was back at work, helping with branding.[8]

The second fortunate choice of employee was Johnny Peirce, who was foreman of the Nowood herds in the Big Horn Basin. In early June, Plunkett took Watson over the mountains and turned him over to Peirce. For five days, he had a pleasant time riding with Watson and going over accounts with the "honest & honorable" Peirce. His consistently complimentary references to these two men were highly unusual; there was not even minimal criticism.[9]

In the Nowood country in June, Plunkett and Peirce were joined by one of the Wyoming Stock Growers Association's stock detectives, those shadowy characters who enforced the livestock laws for the association. Following is Plunkett's description of this man, who is not named: "He was a man of the cool dark devil type one sometimes meets who are quite an interesting study. I knew his history & know that he is a man who carries his life in his hand. A strong cool nerve one would have credited him with, but the careless sleepy manner wh[ich] seemed unaffected didn't look like the energy which his calling required."[10]

In that busy summer of 1885, Boughton fell ill with typhoid fever, and Plunkett decided to go down to the Ione ranch to take over Boughton's duties, telling the latter that he had nothing else to do at the time. "This was a white lie," he confided to his diary. "I never was so busy." Busy or not, when Plunkett went out to visit the ranch, he had no instructions for reaching the headquarters from the Hutton station on the railroad. At the station, he found a pony that was tied to a telegraph pole. The railroad station agent (whose age Plunkett estimated at 15) was of no help, telling him to ride

[8] Plunkett diaries, 8, 9, and 18 May 1885.

[9] On June 9, Plunkett wrote, "I slept with Watson having no bed of my own. He is an excellent bedfellow. Never moves all night. Still I often lose a night's sleep owing to the hardness of the ground. Bony people with wide hips need a mattress." Plunkett diaries, 8, 9, and 13 June 1885.

[10] Ibid., 11 June 1885.

toward a certain mountain peak that could be seen in the distance. Plunkett mounted the pony, still concerned that he would get lost, and only then did he find the answer that had eluded him. "The pony knew the way," he wrote. Boughton slowly recovered, and the strain on Plunkett eased when Boughton's brother arrived to help with the Ione ranch.[11]

Near the end of June, a surveyor arrived to assist in more land and irrigation filings, and Plunkett set down his opinion of American-style professionals, doubtless influenced in part by his unhappy experience with the lawyers in the Searight affair:

> In the afternoon surveyor arrived. He seemed a man of moderate intelligence, but what a miserable lot they are. Why is it that all professional men—doctors, lawyers, clergymen, architects, & c, & c—in the west are so inferior to the unprofessional men? Suppose because the material advantages of this western country enable a man to do better for himself than by working for wages. Perhaps too the majority of western men had the common school education which though far superior to our national education is hardly a first class one.[12]

When Frontier Company opened its cattle feeding operation on a farm at Herman, Nebraska, Plunkett sent Gilchrist there to manage the operations, thus sidestepping the argument over Gilchrist's range management duties. The farm was north of Omaha on the railroad, about six miles from the Missouri River and near an economical supply of corn. When he visited the place in August, Plunkett was pleased. While in Nebraska, he and Gilchrist visited the feeding operation of the Union Cattle Company, which was large enough to feed 3,750 head at a time, as compared with the 1,600-head capacity of Frontier Company's Herman operation.[13]

Back at the ranch on Powder River, range conditions were poor and feed was scarce, although the cattle still looked well. Windsor had gone east to see his family, leaving Old Pete to do the cooking

---

[11] Ibid., 2 and 7 Aug. 1885. William St. Andrew Rouse-Boughton was Edward's older brother and heir to their father's baronetcy.

[12] Ibid., 26 June 1885.

[13] Because the Frontier Co. was short of capital, Plunkett got the seller, V. G. Lantry, to take his proceeds from the land sale in Frontier Co. stock. Ibid., 12 May 1885, 11–14 Aug. 1885, and 12 Sept. 1885.

for assorted passers-by, and everything in the house was so filthy that Plunkett slept outside in a tent. Plunkett went over the mountains to join the fall roundup, which he found on Paint Rock Creek, where Peirce had his wagon with "that very nice boy Watson." After a good night's sleep, he rose before the first light of dawn and rode out for a thirty-mile gallop on the roundup circle. He got fresh information on range conditions as the cowboys on the roundup told him it was "dangerously" overstocked, and he resolved to "lighten up" the herd. When the roundup finished, Plunkett was bitterly disappointed to learn they had found only 500 marketable beef animals instead of the 800 he expected.[14]

Other big ranches were also cutting costs, a move that became industry-wide later, and in early September, Plunkett personally felt the backlash of this policy change. When he arrived at the Pratt and Ferris ranch on Clear Creek at noon, he found Colonel Pratt was not at home, and the workers did not offer him either food or shelter, forcing him to camp out for the night. In late October, the Wyoming Stock Growers Association held a special meeting in Cheyenne to consider what to do about the worsening situation.

The newspapers were told that the association's special meeting was called to recommend amendments to the Maverick Law, but the meeting also was called to adopt cost-cutting measures. Two measures were approved: reducing wages for the cowhands and ending the "free board" custom at ranches. The latter initiative ended the custom whereby the laid-off cowboys could spend the winter on the range riding the grubline from ranch to ranch. Now, when the men were laid off after the fall work was completed, they would have to fend for themselves through the winter until work became available in the spring.[15]

The 1885–86 winter was the first in which the laid-off cowboys no longer rode the grubline, and out on the range the word got around that Horace Plunkett was a driving force behind this

[14] Ibid., 29 Aug. 1885; 3, 4, 12, and 30 Sept. 1885; and 3 Oct. 1885.

[15] The special meeting on 26 Oct. 1885 took a total of 6 hr. Ibid, 26 Oct. 1885. The meeting was attended by forty-two members in person and thirty-one by proxy and was ostensibly to recommend the Maverick Law amendments and to elect a delegation to the national convention in Chicago. None of the newspaper accounts mentioned the cost-cutting measures, which must have been discussed in a secret meeting. *Cheyenne Daily Sun*, 27 Oct. 1885.

change. When he returned to the range the following spring he was concerned the cowboys might go on strike, but that did not happen, although almost everyone was cold toward him. Only Johnny Peirce was cordial, and Plunkett wrote, "It is unpleasant being scowled at & talked at by the blackguards."[16]

The 1885 annual meeting of Frontier Company was held at Herman, and there Plunkett got the shareholders to agree (with John Coble dissenting) to give him full responsibility for range management. After the meeting, Plunkett sent Beau Watson to Herman to take over bookkeeping responsibilities there, so that he now had two trusted employees in place watching over the Frontier Company's financial affairs—Beau Watson in Herman and John Chaplin in Cheyenne. Despite efforts at cost-cutting, cash was still in short supply, and Plunkett and Gilchrist had to personally endorse the Frontier Company's notes with the Nebraska National Bank so that the company could borrow money for its operations.[17]

Plunkett's financial worries were not confined to the tile and ranching businesses, for the Wheatland Project needed money, too. Plunkett played an important role in raising money for the Wyoming Development Company's Wheatland Project, although the details of that activity are unknown; it may have involved arranging British bank loans. In any case, by the fall of 1885, foreign loans were no longer available, and the company had to turn once again to the shareholders for financial help. The shareholders agreed to issue 8 percent bonds, to be purchased by shareholders in proportion to their shareholdings at 65 percent of face value.[18]

Even more work beset Plunkett from all directions as the year progressed. Dick Frewen came to ask his help with the muddled affairs of his Dakota company, and Willy Peters wanted to talk about the problems of the Powder River Company (he had succeeded

---

[16] Plunkett diaries, 18 June 1886.

[17] Coble would have preferred to make Windsor the range foreman. Ibid., 2 and 4 Nov. 1885.

[18] Three shareholders opted not to participate in the bond issue. The nonparticipating shareholders were W. C. Irvine, Plunkett's friend; W. D. Watson-Smyth; and R. Fulton Cutting, one of the investors Sturgis had obtained in New York. Ibid., 23 Oct. 1885; and Wyoming Development Co. minutes, 19 Oct. 1885, Carey collection, American Heritage Center, University of Wyoming (hereafter cited as Wyoming Development Co. minutes).

Moreton Frewen as manager). Then there was the Wyoming Development Company, the Union Cattle Company, and the Bush–Swan Electric Light Company, all with more or less urgent needs. "If I don't go mad during this month, I shall die sane," he said.[19]

The problems of Powder River Company did not have a direct impact on Plunkett during 1885, but they did not go away, either. After Plunkett refused to accept Moreton Frewen's offer for him to manage the company, Frewen hired Willy Peters, paying Peters's salary out of his own pocket. Frewen then got into a long-running dispute with the board over its dividend policy, and at the end of 1885, he sued the board over this matter. He then asked the board to dismiss Peters, and when the board refused the conflict spilled over into the public arena. By the spring of 1886, there was no market at all for the company's preferred shares, and when the shareholders met there were fireworks on both sides. According to one account, "Mud was thrown freely by the directors at the manager, and by the manager at the directors."[20]

At the beginning of March 1886, before Plunkett returned to America, Lord Rosslyn asked him to join Powder River Company's board. Rosslyn was then a member of a special board committee, charged to look into company operations and to cut expenses. Although Rosslyn may not have had the authority to hire Plunkett, the idea was logical. Plunkett's ranch was adjacent to the company's range, and although he was never a shareholder, Plunkett knew the company well.

Although he had toyed with the idea when Moreton Frewen offered him the management, the company's situation had deteriorated in the meantime, and Plunkett refused Rosslyn's offer. Nevertheless, he also wrote a long letter to the secretary of the company in which he stated his own view of the situation. This was a technique he used many times, for he considered it a way to defuse controversy. Unfortunately, Plunkett's "conciliatory" letters typically criticized all sides in a controversy, and generally the effect was to annoy everyone. This case was no exception.[21]

---

[19] Plunkett diaries, 31 July 1885 and 2 and 15 Aug. 1885.

[20] Moreton Frewen to Charles Fitch Kemp, 11 Nov. 1885; Charles Fitch Kemp to Moreton Frewen, 26 Nov. 1885; and William Mackenzie to Lord Rosslyn, 12 March 1886, Frewen collection.

[21] Plunkett diaries, 2 March 1886.

Nevertheless, pressure continued to build within the Powder River Company's board to have Plunkett take over American management of the company, and he clearly knew some such effort was afoot, although details are lacking. Incredibly, Moreton Frewen, who had offered the management job to Plunkett, now declared that he would never consent to the change and would even move to Wyoming to live if necessary to recover his own management rights. Just before Plunkett sailed for New York at the beginning of April, he learned that the Powder River board had unanimously appointed him as manager. In New York, Plunkett learned how strongly the tide had turned against Moreton when Leonard Jerome, Frewen's father-in-law, sent for Plunkett to personally endorse his appointment as manager of Powder River Company.[22]

Plunkett agreed to take the position as manager of Powder River Company, but he refused to take any salary for the work, undoubtedly hoping that this gesture would avoid criticism. If that was his reason, he was very wrong, for he was about to suffer perhaps his most difficult season on the western range. Many furious arguments flew between the Frewen brothers and Plunkett, and at times, these confrontations spilled over into public meetings and into the newspapers.[23]

Although one could hardly find a greater difference in management *style* than existed between Moreton Frewen and Horace Plunkett, a dispassionate examination of their principal *objectives* reveals that they were not that far apart. Nevertheless, in fairness to Plunkett, it is important to note that Moreton Frewen could be an exceptionally difficult person even when he agreed with you. Many of the arguments that swirled in those years dealt with peripheral issues, and after a while, the two sides seemed to be interested only in arguing—not in results.

A major argument arose between the two men over the matter of sending Powder River herds north to Canada to relieve over-

[22] Before leaving England, Plunkett went to see Lord Rosslyn, who was a Frewen partisan, to discuss the situation, but the conversation did not go well. The earl chose to make small talk about the fact that his children's nurse had been the same woman as the Dunsany family had used, and this diversion served only to annoy Plunkett, who declared the earl was "not the perfect gentleman." Ibid., 21 March 1886 and 11 April 1886.

[23] Horace Plunkett to Charles Fitch Kemp, 10 May 1886, Frewen collection. Also, Plunkett diaries, 2 and 11 April 1886

crowding on the Wyoming range. Yet, it was Moreton Frewen who leased the Canadian range in the first place, and he undoubtedly would have moved the cattle if he had remained manager. Although at first Plunkett was not convinced that moving the herds a good idea, in the end he was the one who sent the Powder River herds to Alberta. After taking over management, Plunkett at once put his own stamp on the operation by sending Beau Watson from Herman, Nebraska, to Powder River Company's headquarters. There, Watson would be Horace's familiar—and trusted—eyes and ears on the range. At the beginning of May, Plunkett set aside his other Powder River Company troubles for three days to teach Beau bookkeeping.[24]

Dick Frewen was as furious as his brother over the Plunkett appointment, and he tried to undercut Plunkett's authority with the employees. Plunkett had to give F. R. Lingham, foreman of the Superior, Wisconsin, feeding facility, an ultimatum to choose whom he would work for, and Lingham opted for Plunkett. In the end there were few defections, but confrontations with Dick Frewen finally became so disrupting that Plunkett decreed that in the future they could only communicate in writing. This directive did not stop communication between them, and soon Dick wrote, threatening that if he and Moreton did not recover management rights, it was "war to the knife" with Plunkett.[25]

Nor did the London board make matters easy for Plunkett, as cables to and from London continued almost daily, mostly involving questions and responses generated by Moreton Frewen. Then, in May 1886, Plunkett learned that both Moreton Frewen and his older brother Edward had been appointed to the Powder River Company board. Plunkett fumed that the brothers had got control of the "magnificent property," which "they are no more fit to manage than they are to navigate a ship across the Atlantic." He then predicted, "I shall shortly be relieved of the job."[26]

---

[24] On 1 May 1886, Plunkett wrote, "Spent whole day teaching Beau double entry. He learns well. I don't know a better feeling boy than he is and I shall certainly do all I can to help him along in this country." Plunkett diaries, 30 April 1886 and 1–3 May 1886.

[25] Dick Frewen warned Hesse that he should not obey Plunkett's orders, but Hesse remained loyal to Plunkett, who had a high regard for Hesse. E. W. Murphy was fired for his pro-Frewen stance. Ibid., 1 June 1885 and 21–27 and 29 April 1886.

[26] Ibid., 3 May 1886.

No such dismissal order came, however, and in fact, the erratic Moreton soon cabled Plunkett that he would now *support* Plunkett's management of the company. When this news reached Moreton's brother Dick, he was thoroughly disgusted—this time with his own sibling—and decided to give up the fight with Plunkett. Soon, another letter came from Moreton in which he deplored the squabble between Horace and Dick, blandly asserting that the affair was not of his making.[27]

The Powder River Company was not Plunkett's only problem, for the tile company in Brooklyn needed help as well. The tile company secured some additional money in the summer of 1886. Ivery brought his younger brother into the company and also hired a ceramist from England with tile-making experience. These efforts to improve management were not enough to save the company, which also suffered from a cramped, inefficient plant. Ivery interested John Arbuckle of Arbuckle Brothers coffee fortune to take over the operation, effectively ending Plunkett's venture with ornamental tiles. At the beginning of 1887, Plunkett, Denis Lawless, and Otway Cuffe met for dinner and a "final wail" over the unhappy history of the tile company. A year later, Plunkett stopped in New York and signed away the mortgage on the tile company assets, receiving $16,000 for debentures having a face value of $68,000.[28]

When Plunkett arrived at his own ranch in May 1886, he once again found the house "beastly dirty," with no cook and nothing to eat—no butter, eggs, or fresh meat. But there was much work to be done, and the burden of his correspondence was now very significant: on one day he wrote fourteen letters, taking up thirty pages, and on another day he wrote fifteen. He was now having his own arguments with the Powder River Company board (which included

---

[27] Ibid., 12 May 1886 and 24 June 1886.

[28] Ibid., 2 Sept. 1886; and Townsend, "Development of the Tile Industry," 130. Sidney H. Ivery, 18, came to the United States in 1886, as did Frederick H. Wilde, the ceramist who had worked for an English tile company. John Arbuckle, who conceived of the idea of selling coffee in sealed packages, was the largest importer of coffee to the United States. The tile company did not immediately disappear but was reborn at the beginning of 1889 under the name the International Tile and Trim Company. Financial troubles continued to plague the company, and its annual report at the end of 1891 listed assets of $172,000, which were $5,000 less than the company's debts. *Brooklyn Eagle,* 22 Sept. 1888, 21 Oct. 1888, 31 Jan. 1889, 31 March 1889, and 18 Jan. 1892.

the two Frewens), and the directors were choosing up sides. One director, William Beckett Denison, even suggested that Plunkett appeal to the shareholders over the head of the other directors.[29]

At the beginning of May, Plunkett also got the bad news that his earlier concern for Roach Chapman's character was well founded. The Frontier Company foreman had been arrested for horse stealing in Montana, and Plunkett suspected the real charge might be murder. In short order, the governor of Montana sent a requisition to extradite Chapman, and Plunkett decided to help pay for Chapman's legal defense. This courtesy proved unnecessary, however, because two days later, Chapman escaped from the Buffalo jail by betraying the trust of his friend Sheriff Canton.[30]

On July 13, 1886, Plunkett regretfully took another management action with a man who would come to play a tragic role in another sad story, when he discharged twenty-eight-year-old Nate Champion. He laid off Champion not for misfeasance, but because there was not enough work for both the foreman John Nolan and Champion. Plunkett elected to retain Nolan, who was a better handler of man, although he acknowledged that Champion, who had been running a roundup wagon on the Nowood, was a "good" man. Six years later, Champion's dying testament, written in a house on the KC ranch that was under siege by invading cattlemen, earned him a sort of immortality in the history of the western range cattle industry.[31]

In the summer of 1886, Plunkett learned that George Gordon, his best and oldest cowhand, had broken his leg when his horse fell on him. Plunkett and Maxwell rode over the mountain to help the injured man, who was at a Bar X Bar location on Ten Sleep Creek. A doctor from Buffalo, Wyoming, had already set the leg, which was broken in the upper thigh, and the patient appeared to be in comfort, "of a rude description," according to Plunkett. To help

---

[29] Plunkett diaries, 22 May 1886, 25 and 30 June 1886, and 15 Aug. 1886.

[30] Ibid., 7, 8, 10, and 26 May 1886. Chapman asked Canton for a change of underwear, and the sheriff took the prisoner to his own house to get it. Canton left the prisoner alone for a few minutes, whether inadvertently or by design, and Chapman took advantage of the opportunity by escaping. *Big Horn Sentinel,* 15 May 1886. At the time of the 1900 census, Chapman was married and living in Hanover, Ill.

[31] Plunkett diaries, 14 June 1886 and 13 July 1886.

the patient while away the time were a room full of cowboys, all of them smoking, and two "fast ladies" from the "hog ranch" on Spring Creek.[32]

The confrontation between Moreton Frewen and Plunkett threatened physical violence in June, when Moreton, annoyed at one of Horace's letters, wrote, "It is not wise or fair to write what you would hesitate to say face to face," and Horace quickly responded, "I don't think I am that kind of a coward." Moreton, at 180 pounds or so, clearly outweighed Horace, who was nearly fifty pounds lighter, so it was perhaps fortunate for Plunkett that no physical confrontation occurred.[33]

Plunkett was forced to make a decision on moving the Powder River Company's herds north to Canada when he learned that the Canadian government would impose a duty on cattle imports at the beginning of September. He at once decided to move as many cattle as he could before the duty became effective. Unfortunately, as the cowboys rounded up cattle for the trail herds, Plunkett also learned that the herds were turning up short of the book count, exposing yet another weakness in the balance sheet of the company. Nevertheless, the wisdom of moving the herds to Canada soon became clear, for an infestation of grasshoppers on the Powder River range threatened what grass remained there for the cattle.

On May 24, the first herd was ready to move to the Alberta lease, and by the first of July, they were in Alberta. Plunkett was relieved to learn that the herds had survived the long journey reasonably well and that the Alberta range was good. The second herd was ready to start north before the middle of June, but this time Custer County, Montana, seized the cattle in an attempt to collect a tax on them, and Fred Hesse had to go to Miles City to deal with that problem.[34]

In September, Plunkett set out to visit the Canadian operations. His route took him first to Cheyenne, then to Omaha (where he looked at some lots he and Windsor owned), then to St. Paul,

---

[32] Ibid., 27 and 28 July 1886.

[33] Moreton could become physical at times, for he did strike Dick on one occasion when their disagreements got out of hand. Lord Dunsany tried unsuccessfully to cool the fury between Horace and Moreton. Moreton Frewen to Horace Plunkett, 12 June 1886, and Horace Plunkett to Moreton Frewen, 29 June 1886, Frewen collection.

[34] Plunkett diaries, 11 June 1886 and 9 and 26 July 1886.

where he met Moreton Frewen, who was very cold toward him. From St. Paul, Horace went to Winnipeg, Ontario, then to Calgary, Alta, and thence by coal train to Lethbridge, Alta, where he hired a horse (a "plug") and rode the thirty miles to MacLeod, Alberta. There he learned that 300 cattle had died en route, but the 6,800 that survived were in good condition. On the way back from Canada, Plunkett stopped at Frontier Company's feeding operation at Herman, Nebraska, and learned that the facility had lost $25,000 in less than a year. Back at the Powder River on September 19, Plunkett calculated he had traveled 5,287 miles in twenty-two days, or 240 miles per day.[35]

Plunkett was not finished with rigorous travel. He set out to find the roundup in the Big Horn Basin, and night fell before he could locate the camp, leaving him with only one blanket and no food. Fortunately, he came upon a cowboy and a miner who were en route from Sundance, Wyoming, to Fort Washakie, Wyoming, and he camped with them. The other two men insisted that all blankets be divided equally, because they each had two as well as overcoats. In the morning, he found that he had spent the night only three-quarters of a mile from the roundup camp. Plunkett was not eager to tell the cowhands of his discomfort the night before, but word of his secret had already reached the hands, who asked whether the hotel he had stopped at was on the European or the American plan.[36]

More sleepless nights followed. Two days after his incident at the roundup, on his way back across the mountain, Plunkett stopped at the old Shield ranch. There were only two bunks at the ranch—each four feet wide—for five cowboys. Two men slept on the floor and the other three in the bunks. "My bedfellow stank & turned. I slept not a wink," Plunkett groused. Two weeks later, he and Hesse were back on the mountain, where they slept three in a bed, Plunkett in the middle. The one on the right snored and the one on the left ground his teeth. "It was like going to bed with a blast furnace at one ear and a grist mill at the other," he said.[37]

The financial affairs of Powder River Company were desperate

---

[35] Ibid., 24 Aug.–8, 14, and 19 Sept. 1886. On this trip of 29 days, only seven days did not involve some travel.

[36] Ibid, 22 and 23 Sept. 1886.

[37] Ibid., 25 Sept. 1886 and 10 Oct. 1886.

during Plunkett's entire first year as manager: $30,000 borrowed at the First National of Cheyenne fell due April 2 and had to be extended, and there was also an overdraft at that bank. Plunkett borrowed $3,000 on his own note to provide working capital, and these obligations, together with $10,000 borrowed from a director, added up to a total of $79,000 that had to be raised at once.[38]

Plunkett first approached Morton E. Post, a private banker in Cheyenne who was also a director of Wyoming Development Company, to seek a solution to the cash crisis. After much discussion, Post seemed willing to lend $60,000, secured by a chattel mortgage on the company's properties, but Post insisted that Plunkett also settle Moreton Frewen's personal loan with the Post bank. Plunkett refused that condition and went to see the First National Bank in Cheyenne. The First National Bank offered to advance the money for the summer's operations, provided Plunkett remained as manager of the company and provided that $50,000 to repay the loan was forthcoming "reasonably soon."[39]

These negotiations took some time to complete, and meanwhile the bank began to bounce Powder River Company's checks. Total disaster was only averted when Richard Frewen transferred $22,000 to the company from his own Dakota Stock & Grazing Company, without consulting anyone in the Dakota Company. This was a rare instance when one of Dick's reckless acts actually benefited the Powder River Company for a short time, until the Dakota Company banks recovered its funds.[40]

Moreton Frewen told Plunkett that he had arranged a $30,000 loan in Duluth, Minnesota, to cover at least a part of Powder River Company's requirements. Plunkett went to Duluth to collect the proceeds of this supposed loan, and there he found the facts were very different from Moreton's story. Not only were the bankers unprepared to lend the money, but they were also critical of the company's unbusinesslike methods. Nevertheless, Plunkett was finally able to negotiate a new loan from the Duluth bankers, but for only $15,000.

---

[38] Horace Plunkett to Charles Fitch Kemp, 7 May 1886, Frewen collection.

[39] Ibid., 13 and 14 May 1886, Frewen collection.

[40] Charles Fitch Kemp to Moreton Frewen, 19 May 1886; Horace Plunkett to Charles Fitch Kemp, 16 June 1886; Richard Frewen to Moreton Frewen, 17 June 1886; and Richard Frewen to Charles W. Middleton Kemp, 17 July 1886, Frewen collection.

Plunkett had assumed that he needed to raise only enough money to keep the American operations of the company afloat, but when the London office learned of Plunkett's success in raising money, they asked him to send £5,000 over to London, which would have left only £7,000 for use in America. In the end, money had to flow from London to America, and £15,000 finally arrived in December 1886. This cash infusion bought some time, and Plunkett was determined to use that time to liquidate the company in an orderly fashion, even though the London board had not yet agreed to that course of action.[41]

In these months before the disastrous winter of 1886–87, Plunkett also began to worry about the future of his own Frontier Company, and he and Gilchrist cast about for ways to extricate themselves and their backers from it. The shareholders decided to partially liquidate Frontier Company by dividing the several outlying properties among themselves. Plunkett expected that the major shareholders would engage in a sort of auction to determine values for the individual properties, but that he would take a proportionate interest in each of them so that he could not be accused of taking the best properties for himself.[42]

Windsor, Coble, Alexis Roche, Gilchrist, Boughton, and Plunkett met in Omaha near the end of October to divide up the outlying properties of Frontier Company. Windsor, Coble, and Plunkett took the Herman facility in equal shares in the expectation that it could be sold, and Windsor and Coble took the Iron Mountain ranch. The Crow Creek ranch was the best asset in the company and was the most controversial item. Gilchrist had brought the ranch into the company, and he insisted that he should have the right to buy it back at his price, without the need to meet other bids. In the end, he got the Crow Creek ranch, but only after Boughton and Plunkett carried the price $12,000 over his initial bid.[43]

The shrunken Frontier Company still had its cattle operation and $80,000 in debts, and now that Gilchrist had left the company, it was necessary to tell the Nebraska National Bank about the reor-

---

[41] Plunkett diaries, 15 Nov. 1886.

[42] Ibid., 30 Aug. 1886 and 4 Oct. 1886.

[43] Ibid., 18, 20, and 27–30 Oct. 1886.

ganization of the corporation, for Gilchrist had endorsed the company notes there. Unfortunately, when Plunkett told Henry W. Yates, the bank president, about the corporation reorganization, Yates declined to continue to loan money to company.[44]

Boughton was still a substantial shareholder of Frontier Company, and Chaplin asked him to endorse the company's note at the Stock Growers National Bank in Cheyenne. Boughton flatly refused, saying that to do so would tie up his other business (presumably chiefly the Ione Company). Then, Boughton added, "I cannot in justice to my own people put my name to notes of this amount for indefinite time," and he declared, "This is a bona fide crisis." He warned Plunkett that if he left Wyoming before the matter was straightened out, "The [Frontier] Co. must very shortly, in fact, at once, go under." Plunkett did not comment on this letter in his diary, but he did finally eliminate this crisis by arranging financing.[45]

A natural disaster added another threat to Frontier Company's financial problems. At the beginning of November, a huge prairie fire struck at Tekamah, north of the Herman farm, and swept over the Herman lands, threatening hay stacks and fences and scattering the cattle. Fortunately, after the fire was out, Plunkett learned that the company's loss was not more than $2,000, for the barns and most of the hay had been saved, although a bridge was burned.[46]

Although the Herman facility was still for sale, Plunkett sent some of Powder River Company's cattle to Herman to be kept during the 1886–87 winter, for which the partners were to be paid for weight added. Unfortunately, when he returned to Herman the following spring, he learned that the winter's operations had again lost money.[47]

Plunkett also recognized that the Tebbetts venture must be sold, at a sacrifice if necessary. The following fall, after a further loss of some £600, Plunkett sold his share of the enterprise, including his

---

[44] Ibid., 3 Nov. 1886.

[45] E. S. R. Boughton to Horace Plunkett, 13 Nov. 1886, Ione letterpress. It is possible that this flare-up led to Plunkett's decision to eliminate Boughton from the list of those with whom he ultimately shared his "surplus" when the company was finally wound up.

[46] Plunkett diaries, 5 Nov. 1886.

[47] Ibid., 30 April 1887.

note of nearly $4,000 from the partnership, to Charlie Wyndham-Quin. The transaction apparently involved no cash, but young Quin assigned his shares in Frontier Company to Plunkett together with a $5,000 receivable from that company.[48]

A final loss to Plunkett's Powder River ranching operation occurred in October 1886 with the death of Old Paddy, the greyhound Alexis brought from England in 1881. A real survivor, the dog had been badly poisoned once, torn by wolves and badgers, and scalded by prickly pears. He was "the bravest of dogs," and although his fighting days were "full of adversity," he was also "most amiable." He was a full member of the ranch family, and all mourned his passing.[49]

Plunkett's year in America ended on December 10, when he boarded the *Servia* after dining in New York with the Marquis de Mores, who was also having financial troubles in the western ranching business in Dakota Territory. It had been an awful year, full of labors, with few rewards, and more trouble was on the way, for the notorious winter of 1886–87 had only just begun out on the western ranges.[50]

On his way back to Europe in December 1886, Plunkett wrote his report to the shareholders of Powder River Company. In it he said, "I am shortly about to resign the management of the Powder River Cattle Company, because I find it a position compared with which the throne of Bulgaria is a bed of roses."[51]

He kept his promise the following February in a meeting with Powder River Company's board, where he resigned. All those present, except Edward Frewen, asked him to reconsider, but he refused. Privately, Horace told William Mackenzie that he would take the job if he could have absolute control, was paid £600 for the year just past, was paid £600 for the current year, and if he could hire Frank Kemp as his assistant at £400. A month later, perhaps

[48] In analyzing the reason for the partnership losses, Plunkett said that Tibbets did not realize that he had neither a breeding establishment, nor a free range horse business, but a mixture of the two. Ibid., 28 Aug. 1884, 4 Oct. 1886, and 1 and 2 Oct. 1887.

[49] Ibid., 6 Oct. 1886.

[50] Ibid., 10 Dec. 1886.

[51] The throne of Bulgaria was vacant at the time Plunkett wrote his report, Prince Alexander of Battenberg having been forced to abdicate in August 1886. The Bulgarian parliament finally proclaimed Ferdinand of Saxe–Coburg and Gotha as prince regnant in July 1887, without waiting for the approval of the great powers of Europe.

in reaction to the news from the western range, Horace added an additional condition to his proposal: he would only manage the company to liquidate it.[52]

A few days later, another wild card entered this incredibly muddled game. Sir John Lister-Kaye approached Jim Winn with a proposal to launch a large land and cattle company, which he wanted Plunkett to manage. The cornerstone of this company was to be Winn's Big Horn Cattle Company and Powder River Company, and Lister-Kaye assured Winn that he had the backing of the government of Canada and the Canadian Pacific railroad.[53]

Plunkett at once went to meet Lister-Kaye, where he sharply criticized the cattle portion of the scheme, saying, "It was wild." Afterward, he went to see Winn's father, Lord St. Oswald, who was of the same opinion but who hoped the government and railroad backing might produce something of value. Lord Dunsany was also eager to have Horace leave the range cattle business, saying, "The day that you give up America will be the happiest day of my life." Horace continued to hope that Lister-Kaye would buy the two companies.[54]

The annual meeting of Powder River Company was held on March 29, 1887, and was attended by about thirty shareholders. Lord St. Oswald was in the chair, trying without success to maintain at least a modicum of decorum. Moreton Frewen (who may have had some news of winter losses out on the Wyoming range) attacked Plunkett, saying,

> I know Mr. Plunkett well; he has been my neighbour on Powder for many years. A more unfortunate choice [as a manager], as I wrote at the time, could not have been made. He is altogether too weak a man, in my judgment, for the critical times on which we had fallen in this business. The result, anyhow, is that, left to shift for themselves on an absolutely exhausted range, our property is at this moment squandered, dead or dying along the hill sides or streams of the Big Horn Mountains.

---

[52] Those present at the London meeting were Edward Frewen, Joseph Richardson, William Mackenzie, and George D. Stibbard. Plunkett diaries, 15 Feb. 1887 and 18 March 1887.

[53] Sir John Pepsy Lister-Kaye, who succeeded to the baronetcy in 1871, was a fellow classmate of Plunkett at Eton. He courted Clara Jerome before Moreton Frewen appeared on the scene, and in 1881 he married Natica Yznaga, the sister of the Duchess of Manchester.

[54] Plunkett diaries, 22–25 March 1887.

No one chose to answer Frewen, but after the meeting some of the shareholders pressed Plunkett to resume management. He said he would give them a written proposal.[55]

For Powder River Company, losses on the range were only a part of the problems that beset Plunkett's efforts to liquidate the company. Moreton Frewen urged Plunkett's father to use his influence to get Horace to resign, and in Wyoming, Frewen filed a lawsuit, claiming he owned the lands that he had filed on for the company. In a circular addressed to shareholders, Frewen again attacked Horace, claiming that the latter had never seen a winter storm on the prairies and had as little conception of what happened in the winter "as the Grand Lama."[56]

The ordinary shareholders of the company had long since resigned themselves to the loss of their investment, because the assets of the company were less than the preferred shareholders' claims. Plunkett estimated that he could realize a total of only £75,000 to pay the preferred shareholder claims of some £133,000, including accrued interest.[57]

Plunkett was aware of the increased risks to open range cattle ranching from overgrazing well before the winter of 1886–87. Indeed, it can be said that Powder River Company was headed toward liquidation before that winter, and Plunkett's restructuring of Frontier Company also reflected the need to deal with poorer conditions. Nevertheless, the awesome losses of that winter enormously complicated any effort to conduct an orderly exit from the business.

---

[55] *Financial News*, 1 April 1887; and Plunkett diaries, 29 March 1887.

[56] Moreton Frewen to Lord Dunsany, 5 April 1887; Lord Dunsany to Moreton Frewen, 29 April 1887; and "Statement to the Shareholders of the Powder River Cattle Company," 24 May 1887, Frewen collection. Also, Plunkett diaries, 18 April 1887.

[57] Plunkett diaries, 8 Aug. 1887.

7

A Bad Winter and More Bad News

LOSSES FROM THE WINTER of 1886–87 not only doomed the open range cattle industry, but they also destabilized the entire western cattle industry and forced the big cattlemen to try drastic ways to restore profitability. In his role as manager of Powder River Company and as president of his own Frontier Company, Plunkett was caught up in these efforts.

The winter of 1886–87 is legendary in the history of the range cattle industry on the High Plains and in Texas. Descriptions of cattle ranches in this period are replete with references to losses, some of them so obviously exaggerated as to merit little attention but, in total, telling the story of a very considerable loss on the range. During the winter, western editors were reluctant to pass on negative information, so that while the storms were raging there was little information about the weather in many of the major papers. Most solid information about the winter came after the fact, when the shrunken calf brand totals in the spring of 1887 told of dead cows, and the number of steers gathered for market in the fall fell far below expectations. Clearly the surviving herds were smaller than they should have been, although critics have argued that the winter was unfairly blamed for losses from other causes, including cattle bought on book count that never existed in the first place.

In 1942, Dr. Taft Alfred Larson wrote an article on the famous winter that was based on a comprehensive review of the sources, and from this article one can piece together a general view of conditions. The summer of 1886 was dry, and the range, which was unquestionably overcrowded in many parts of Wyoming, was in poor condition when the winter struck in November. There are conflicting stories about the severity of the early storms, but there is general agreement that the persistent winds drifted the snow badly. The surface of the drifts was frozen hard enough for the stagecoach to drive over them, and the starving cattle could not reach the grass frozen below. Moreover, after winter commenced in November there was no significant break in the procession of storms sweeping across the plains—no warm days to thaw the frozen landscape. Herds suffered terribly, and even some ranch hands froze to death, while others narrowly escaped the same fate.[1]

There is a contemporary account of life on the northern range during the early part of that winter, for Henry Maxwell was at Plunkett's EK ranch for several months in the late fall of 1886. When the first winter storms hit the Powder River Basin, the fall roundup was still out on the range, and on November 10, Maxwell wrote that storms were coming "every other day." Already there was no feed for the cattle.[2]

Then a "most fearfully heavy snow storm" laid down more snow than the entire previous winter and halted the roundup. The returning cowhands brought firsthand reports of grass covered by six to seven inches of snow, and all of the available water was locked below the ice of frozen rivers. The horses had gone several days without feed, and one from the 76 ranch died before it reached home. The rest of the horses were in no condition to drive the cattle to the railroad, and the snow cover on the range made it unlikely that the herd could be moved anyway. The steers were not especially fat, and Maxwell feared that even those that survived the drive to the railroad would be too poor to sell. Because he was alone at the ranch, he had to make the key decisions, for the idle crew would draw wages until they were laid off. Maxwell turned the steers into the pasture at the ranch to fend for themselves and paid off the men. Then he left the range, ending this eyewitness record of that fearsome winter.[3]

---

[1] Taft Alfred Larson, "Winter of 1886–87 in Wyoming."

[2] Henry Maxwell to John Chaplin, 10 Nov. 1886, Fred G. S. Hesse collection, American Heritage Center, University of Wyoming (hereafter cited as Hesse letterbook).

[3] Henry Maxwell to Alexis Roche, 20 Nov. 1886; and Henry Maxwell to Horace Plunkett, 20 Nov. 1886, Hesse letterbook. Also, Plunkett diaries, 11 March 1887.

Down in Cheyenne, the first reliable news of the range's condition came to the city not through the newspapers but on foot: hungry cattle looking for food appeared on the outskirts of the city by the first week of January, and by March, the police gave up trying to keep them out of the city itself. Plunkett's first diary entry regarding the winter was one terse sentence in March when he was still overseas, "News from the Ranch very bad 30 to 40 per cent loss expected."[4]

On April 16 Plunkett arrived in New York, where he learned of the desperate efforts of the big ranchers to salvage some value from their depleted properties by trying to control the marketing of beef. Two schemes to accomplish this purpose were floated in 1887. The first was a plan to organize a giant cattle trust, a $100 million monopoly of cattle producers in combination with the Nelson Morris packing plant in Chicago. The idea for the trust, which Plunkett learned about from Tom Sturgis, apparently came from a speech by Edward M. McGillin, a man heavily invested in cattle.[5]

This was a period of great consolidation in American business, and a number of combinations, or "trusts," were formed to dominate the industries in which they operated. Several of them are well known, and the government broke up some, including the Standard Oil Trust. The crippled cattle industry now hoped their own trust could control the supply of beef and its price, and McGillin persuaded Sturgis to lead the effort to form the trust. On May 3, 1887, the *New York Times* reported the formation of the trust, which was modeled after the Standard Oil Trust.[6]

The trust was not short of big names. The chairman was Thomas Sturgis, who had just stepped down as secretary of the Wyoming Stock Growers Association, and the vice-chairman was Charles F.

---

[4] *Cheyenne Daily Sun*, 12 Feb. 1887; and Plunkett diaries, 11 March 1887.

[5] The McGillin brothers were Cleveland dry goods merchants who purchased the T J C Ranch on the Republican River in 1884 and incorporated the Harlem Cattle Co. The company was managed by W. J. McGillin, with Edward M. McGillin as president. Yost, *Call of the Range*, 131.

[6] *New York Times*, 3 May 1887. The Standard Oil Trust agreement, drafted in January 1882, established an organization to own Standard Oil Company of Ohio and forty other Rockefeller companies. Tom Sturgis's younger brother, Frank Knight Sturgis, bragged that he had a copy of the Standard Oil Trust agreement. Gene M. Gressley wrote an article describing the American Cattle Trust in the *Pacific Historical Review* 30 (February 1961), 61–77, but elsewhere the organization has generally been ignored.

Smillie, an importer in New York who had been president of T. B. Hord's Lance Creek Cattle Company of Lusk, Wyoming. Charles T. Leonhardt, a New York stockbroker, was secretary–treasurer.[7]

Range management was the responsibility of general manager Richard G. Head, who was former manager of the huge Prairie Land and Cattle Company; he also managed the New Mexico department of the trust. The Wyoming department was headed by former Wyoming governor Francis E. Warren, and the Colorado department was headed by Jared L. Brush, sometime partner of Colorado governor John L. Routt, who was a director of the trust. Captain John Thomas Lytle, who had organized cattle drives from Texas to the Kansas railheads, managed the Texas department.

The board of directors had representation from both western cattlemen and eastern financial figures. The cattlemen were Sturgis, Warren, Head, Routt, Lytle, and Christopher Columbus Slaughter of Texas. Board members from the East included Smillie, Leonhardt, New York broker Richard T. Wilson, Jr., and ironmonger Samuel Thomas (who was also the controlling shareholder of the Chase National Bank). Of more than incidental importance was the presence on the board of Charles McGhee, who was the father-in-law of George W. Baxter, a Wyoming rancher and former governor of Wyoming Territory. The eastern directors proved to be troublesome partners in the operation, as they wanted to maximize cash flow by selling off all marketable cattle at each shipping season, rather than conducting large feeding operations.[8]

The trust was designed to eliminate the commission men from the sales chain and thereby capture the 50¢ or so per animal that those men would otherwise earn. To do this, the trust had to go into the fattening and packing businesses, purchasing an idle plant from Nelson Morris in Chicago (for $2 million) and feeding farms at Gilmore, Nebraska. The *New York Times* reported that Morris received $2 a head to kill and market the beef from the trust, and he also was named "advisor" to the directors. The valuation for the

[7] The trust was originally headed by Edward McGillin, but he was soon removed and was succeeded by Sturgis. Yost, *Call of the Range*, 150.

[8] The western cattlemen on the board were Gov. John L. Routt of Colorado, Richard G. Head of New Mexico, C. C. Slaughter and John T. Lytle of Texas, and Thomas Sturgis and Francis E. Warren of Wyoming. Gressley, *Bankers and Cattlemen*, 261–62. Baxter succeeded Warren after the latter resigned from the trust.

Morris plant was supposedly at the same 25 percent ratio to real value as for the cattle herds and ranches. Actually, Morris's personal role in the trust was apparently minor, and the advice he gave was often ignored, but his connection with the trust only served to heighten the suspicion of critics that this combination was allied with the packers. In fact, the trust was intended to be a competitor of the packers.[9]

Cattle ranchers who transferred their herds to the trust received trust certificates with a face value of 25 percent of the value of the herds, and these certificates traded on the stock market, just as the Standard Oil Trust certificates were traded. The owner could sell his certificates immediately or retain them, hoping to sell when their value increased.[10]

While the cattle trust moved forward with its organization, a competing group of cattlemen in Denver in the fall of 1887 proposed forming a "beef pool." This organization was another effort to cut the commission merchants out of the marketing chain by marketing western cattle in combination with Armour Company. The secretary of the Colorado Cattle Growers Association sent Asa Shinn Mercer to Chicago to try to negotiate the American Beef Pool with Phil Armour, who was to be the pool's exclusive purchaser of beef cattle. Armour Company signed the proposed contract, which would have paid Armour a flat $2.50 per head to kill cattle and market the meat. Like the American Cattle Trust, this concept faced long odds, including the opposition from the trust, commission merchants, and Armour's competitors in the packing business, and it never attracted enough support among cattlemen to get off the ground.[11]

Plunkett liked the idea of the cattle trust, and at his urging the other significant shareholders of Frontier Company agreed to

[9] Nelson Morris was born in Bavaria in 1839 and came to Chicago at the age of fifteen. He invested in large ranches in addition to his packing operation. The "Big Four" packers in Chicago were a somewhat variable group, originally consisting of Armour, Swift, Morris, and G. H. Hammond; after Hammond was acquired by Armour, the fourth member of the group was Swarzchild and Sulzberger. In other packing centers, Cudahy was regarded as the fourth packer, instead of Hammond. To complete the integration of the trust's business, contracts for canned beef were also negotiated with the French and Belgian governments. *New-York Times*, 4 May 1887 and 23 May 1888.

[10] Gressley, *Bankers and Cattlemen*, 263; and Plunkett diaries, 17 and 22 April 1887.

[11] Leech and Carroll, *Armour and His Times*, 191.

commit the company's operations to the trust. He also advised Jim Winn to join the trust. Although there were many details to negotiate, these decisions and the continuing liquidation of Powder River Company signaled the end of an era. The ranches launched in 1879—with high hopes of quick riches from the free range on the public domain—were no more.[12]

On his way west that awful spring, Plunkett stopped at Herman, where the cattle were all gone. He spent a quiet Sunday with Beau Watson before pushing on to Cheyenne and then north to the old ranches, this time using a new route that testified to the advancing settlement of the country. The Northwestern Railroad had built into Wyoming, so that the preferred route was by stage from Cheyenne to Chugwater and to Lusk, then on the Northwestern to Douglas, and finally by stage again to the ranches. Beau Watson came in by train from Herman on the Northwestern to Douglas to join Horace on the second stage leg of the trip.[13]

At the ranch, Plunkett found the place "going to the dogs," as was usually the case when he returned in the spring. To make matters worse, Alexis Roche had thoroughly soured relations with the neighbors, making it impossible to share work with the other ranches. "The country was getting too hot for us all," Plunkett wrote. Poor Alexis had nearly bankrupted his own people, had lost all hope for the future, and even wished he were dead. To avoid further problems, Plunkett fired Alexis, following a difficult confrontation. Their parting distressed him but did not soften his analysis of Alexis: "His disregard of truth, want of education & above all never having gone to a public school but having lorded it over stable boys when he ought to have been fagged & kicked, has spoiled a man of great natural ability."[14] It was a sad ending for a business relationship of eight years that began in 1879 when both men were in their twenties.

The threat of a new crisis loomed for Plunkett on May 24 when he learned that Alex Swan's empire had toppled. At first, Plunkett assumed the Swan insolvency involved the Swan Company itself, rather than Alex Swan and his brother (as was the case). He imme-

---

[12] Plunkett diaries, 9 and 10 May 1887.

[13] Ibid., 30 April 1887; 1, 2, 11, 13, and 14 May 1887.

[14] Ibid., 9, 17, 19, 21, and 22 May 1887.

diately fretted that he might inherit the liability on the railroad lands he had sold to the Swan Company. Fortunately, this crisis did not materialize.

Out on the Powder River range, the full extent of the ravages of the previous winter began to become apparent. Plunkett found the herds were "melting away" and estimated that Powder River Company's herds still remaining in Wyoming and Montana (on Hanging Woman Creek) could be as few as 14,000, excluding calves. In July, Fred Hesse gave Plunkett his final estimate of the losses from the previous winter: 10 percent of the steers and 75 percent of the rest of the herd. These sobering facts added even more substance to Plunkett's determination to liquidate Powder River Company as soon as possible.[15]

The sad story of another young aristocrat who had hoped to make his way in the American West came to an end in May—but the hard winter was not to blame this time. Plunkett learned of the suicide in England of Harry Thynne, the son of a younger son, whose paternal grandfather was the Marquess of Bath and whose maternal grandfather was the Duke of Somerset. When Harry was "cast off" by his parents and had to fend for himself, Moreton Frewen brought him out to work for Powder River Company in 1882, when he was twenty-two. He worked there the next year as well, but life on the range was not for him, in large part because he spent his earnings on whiskey, as Plunkett noted. Returning to England, Harry tried to support himself by his writing, and when he completely ran out of money, he shot himself.[16]

Plunkett spent early June trying to sell Powder River Company's herds, trying to transfer his own herds into the cattle trust, and advising Peters that he and Alston should do the same. A bright spot was Beau Watson, who was at the NH ranch and was "thor-

---

[15] Powder River Co.'s accounts for 1887 estimated the herds (including those in Alberta) at 17,805. Woods, *Moreton Frewen*, 188; and Plunkett diaries, 28 May 1887 and 2 July 1887.

[16] Plunkett said that Henry Boteville Thynne was the "bibulous son of a bibulous father." Plunkett diaries, 31 May 1887. An ironic twist to the story of young Harry Thynne's suicide involved an annuity of $50,000 per year. His cousin, the primary legatee, was due to inherit the annuity when he reached his majority, and Harry was the residual beneficiary. Because the cousin was a healthy nineteen-year-old at the time of Harry's suicide, Harry despaired of ever receiving the annuity, but only eleven days after Harry's death, his cousin's horse fell on him and killed him; thus, Harry would have inherited the annuity had he lived.

oughly in touch with the business" after Plunkett gave him a complete briefing on how he wanted his business affairs conducted and sent him back to Herman and to Omaha. Horace was pleased that by thus using Beau as his alter ego, he could effectively be in two places at once.[17]

In mid-July, the American Cattle Trust sent Luke Voorhees to contract for the transfer of herds to the trust, and these negotiations did not go as smoothly as the ranchers had hoped. Rumors were afoot that the trust showed favoritism in buying the Texas ranches, the Nelson Morris plant in Chicago, and the holdings of Sturgis's Union Cattle Company. The ranchers were consequently wary of Voorhees's efforts. Nevertheless, Plunkett was still optimistic that the combination could work, and Jim Winn also arranged to transfer the herds of his Big Horn Cattle Company to the trust.[18]

A minor controversy flared up between Plunkett and Boughton in the summer of 1887 over the sale of 55 bulls from Ione Company to Powder River Company. Four of the bulls were left behind because they were lame, and Plunkett asked Boughton to take responsibility for the lame bulls. In a "My dear Horace" letter, Boughton rather archly refused, saying, "Business is business, so please send the note [in payment] right away, as per our understanding."[19]

Beginning in the middle of August 1887, Plunkett made another grueling trip to Canada to see how Powder River Company's cattle were faring on the Mosquito Creek range. On the way, he stopped in Omaha to meet Windsor and Beau Watson and then went to the company's operation at Superior, where he suffered another bout with diarrhea. Then he went on to Calgary and from thence to Mosquito Creek to see the company's herds there. The good news was that the cattle were in excellent shape "& mostly there," but Mosquito Creek lived up to its name and there was a cloud of mosquitoes where he camped. This journey of 17 days covered 2,620 miles by rail, 84 by buggy, and 190 on horseback.[20]

Early in September, Plunkett sold the Powder River Company's

---

[17] Plunkett diaries, 7 July 1887.

[18] Plunkett diaries, 15–18 July 1887.

[19] E. S. R. Boughton to Horace C. Plunkett, 21 July 1887; and E. S. R. Boughton to H. S. Oliver, 13 Aug. 1887, Ione letterpress.

[20] Plunkett diaries, 2–29 Aug. 1887.

Superior properties, which were losing £7,000 (chiefly because of Moreton Frewen's interference the previous fall). F. R. Lingham, the company manager at Superior, bought the property for $50,000, but Frewen threatened to stop the sale because some of the lands were still in his name. Powder River Company had to obtain a court order to force Frewen to execute deeds in favor of the company so that the sale could be completed.[21]

Everywhere on the Powder River, people were preparing to leave now that the cattle trust had taken over the herds, and closing down was especially difficult for some. After Peters and Alston transferred their herds to the trust, Walter C. Alston "lost his way," even talking of suicide, whereas Charlie Wyndham-Quin said he was sick and didn't want to ride in the roundup. Horace also closed up the EK headquarters, burning papers "by the bushel," giving away his personal kit (except for one complete outfit in case he came back again), and leaving his pet horse with Booth. Young Millais photographed the horses and wagons at the ranch as a memento of his experiences there.[22]

In Cheyenne, Plunkett's impending departure gave rise to an outpouring of warm feelings toward him. First there was a "rather merry" impromptu dinner, and then thirty members of the Cheyenne Club gave him a formal farewell dinner. The speeches testified to his "high honor" in business, but he was pleasantly surprised to hear particular appreciation for his readiness to help others. He mused, "Many have had my good will & all the help I could give them. That's true."[23]

One of Powder River Company's remaining properties was the

---

[21] The buyer offered a commission of $1,000 to Plunkett, who refused to take it and arranged to have it offered to Lingham, but the latter held out for $1,500. After Frewen capitulated, Plunkett gave a $750 commission to Lingham, rather than the $1,000 he had previously offered. Unfortunately, Lingham had the last word in this negotiation, for when Plunkett told him to draw the money the company owed him, he took $541, consisting of expenses Plunkett called "most iniquitous." Plunkett diaries, 27 April 1887; 21 and 24 June 1887; and 5, 6, and 8 Sept. 1887.

[22] Plunkett settled up with Wyndham-Quin by exchanging his own shares in the money-losing Tebbetts & Co. for Charlie's receivable and shares in Frontier Co. Plunkett diaries, 1 and 19 Oct. 1887. Geoffroy Millais came to the American West as a younger son to make his fortune, not expecting to inherit from his father. However, after he returned home, his nephew died without issue in 1920, and Geoffroy inherited his father's baronetcy, becoming Sir Geoffroy Millais. He died 7 Nov. 1941.

[23] Plunkett diaries, 24 Oct.–2 Nov. 1887.

"natural refrigeration" facility Moreton Frewen built along the Union Pacific at the top of Sherman Hill, between Cheyenne and Laramie. Frewen hoped to butcher cattle there and have the meat "naturally" cooled in the frigid climate until it could be shipped east on the railroad in refrigerator cars. A financial disaster, the facility could not be competitive with the product coming to market from the established packers in Chicago. Moreton tried to sell it without success, and Plunkett finally sold the facility as scrap lumber to a dealer. The facility was located across the Union Pacific railroad from the Ames monument, and Plunkett mused about the two locations. "Funny," he said, "the two monuments on the Rocky Mts. at Sherman, one on south side of track to Oakes Ames of Credit Mobilier fame & on the other side the monument of Frewen's folly & the British investor's gullibility."[24]

Plunkett cut wages at Powder River Company at the end of 1887, reducing cowboys to $31.50 per month and cutting Hesse's annual salary by £120, or about 20 percent. Hesse complained that it was hard to have his pay cut while doing the same work, and Plunkett restored the pay cuts in the spring of 1888 when he increased salaries to $40 per month for experienced hands and to $35 for new hires. Fortunately, the 1887–88 winter was mild, and the remaining cattle came through it in fine condition, yielding a calf brand of 1,719 in the spring of 1888.[25]

---

[24] Ibid., 6 Nov. 1887. The Marquis de Mores constructed a slaughterhouse at Medora, Dakota Territory, that was expected to serve the New York market, where he even acquired retail outlets. This venture, which had a better marketing plan than Frewen's, also failed, for the same marketing reasons that afflicted Frewen's idea. The Ames monument was erected by the Union Pacific to honor Oliver and Oakes Ames for their service to the railroad. Both men were also involved with Credit Mobilier of America, an important and highly profitable construction contractor for the Union Pacific. When Congressman Oakes Ames was head of the company, he sold shares of Credit Mobilier to members of Congress at bargain prices, and he was subsequently censured. The Ames monument was erected at the highest point on the railroad, but the line was subsequently relocated about three miles south, and the railroad donated the monument to the State of Wyoming in 1983.

[25] Fred Hesse was drawing a salary of $3,000 in 1885, according to his interview in the Bancroft collection, American Heritage Center, University of Wyoming. Fred G. S. Hesse to his sister Jenny, 10 Nov. 1887; Fred G. S. Hesse to William J. Mackenzie, 26 Dec. 1887; Fred G. S. Hesse to Charles Carter, 28 Dec. 1887; Fred G. S. Hesse to Jim Winn, 30 Dec. 1887; and Fred G. S. Hesse to Charles Carter, 15 April 1888, Brayer collection, American Heritage Center, University of Wyoming. At the end of 1888, Hesse was drawing $150 per month, but that salary probably reflected reduced duties, as the company was nearing liquidation. Fred G. S. Hesse to Horace Plunkett, 25 July 1889, Hesse letterbook.

Ownership of the Wyoming Development Company was also changing, forcing the company to finance its development outside the shareholder group. The first change occurred when Warren withdrew in 1884 and Thomas Sturgis joined the company. The next year, Sturgis brought in three large shareholders from New York for a total of 4,500 shares, or over 18 percent of the total outstanding. Then in the fall of 1887, Morton Post's bank failed, and his shares passed to other outsiders. A number of the new shareholders were passive investors who did not expect to make further investments in the company, so it was no longer possible to finance Wyoming Development by regular pro rata calls on the shareholders.[26]

In Cheyenne, Plunkett stopped to see two casualties of the cattle business, Willy Peters and his wife, who had opened a boarding house. After his return from the trip to Sherman Hill, Plunkett and nine others went to Mrs. Peters's house for dinner, paying $2 per plate and taking the food away with them. Two days later, Horace hosted a dinner for twelve at her house to let her earn some more money. When he left on the train for the East, a dozen men came to the station to see him off, and he suffered "a slight attack of 'parting-itis.'" He worried most about Alston, who was now in Cheyenne and lacked the funds to go home.[27]

Windsor finally sold the Herman, Nebraska, facility to packer Nelson Morris for $24,000 in cash after Plunkett unsuccessfully tried to get John Clay, Jr., to buy or lease it. The partners Windsor and Plunkett had given up Frontier Company stock nominally worth $26,000 to acquire the facility from Frontier Company, but the cash sale to Morris was one happy ending to an otherwise dismal story. The partners excluded the Hiland property from the sale as well as some other assets worth about $18,000. Horace gave Beau Watson the responsibility of accounting for what was left of Powder River Company, staying a day longer at Herman to make sure that Beau was "thoroughly posted" in that work.[28]

In New York on his way home in November, Plunkett learned Sturgis told Gilchrist that when he transferred the Union Cattle

---

[26] Only 2,500 of the New York shares represented new money, as Sturgis lightened his own holdings at the same time.

[27] Plunkett diaries, 6, 8, and 10 Nov. 1887.

[28] Ibid., 11 and 12 Sept. 1887, 4 Oct. 1887, and 12–16 Nov. 1887.

Company herds to the cattle trust he had protected the Union Cattle Company's notes to Frontier Company. This news greatly relieved one of Plunkett's concerns. However, when he and Gilchrist met Sturgis, Warren, and Lane to discuss Frontier Company's contract with the trust, there was a fight.

The favorable verbal assurances Sturgis had given Plunkett were not a part of the written contract, and when challenged, Sturgis "wiggled" out of the verbal portion. The argument continued until Plunkett had to break off to catch the ship, which he caught only because the captain held the vessel five minutes for him. The next chapter in the Sturgis story came in early February, when Plunkett learned that Sturgis's Union Cattle Company had failed. Plunkett persuaded Windsor to relieve him of liabilities in dealings with that company, in exchange for a payment of $2,000 and the transfer of his share in the WP herds.[29]

Back in England at the beginning of 1888, Plunkett spent a good deal of time with his former colleagues on the range, many of whom were suffering financially in the aftermath of the cattle industry collapse. Denis Lawless, one of the investors, talked of going away to "some portion of the earth" for seven or eight years.[30]

Nevertheless, all was not financial gloom, for Horace still had an appetite for American investments. In mid-March, Frank Kemp cabled to suggest that he invest £500 in Salt Lake City, which was then enjoying a boom period. Plunkett visited Willie Blacker to tell him what had happened to the money he had invested in the cattle business and to reassure him about repayment. In the course of the conversation, Horace also told Blacker about Salt Lake City, and the latter then invested £250 in the Utah project and loaned Plunkett £250 to invest as well.[31]

In a meeting with Sir John Lister-Kaye, Plunkett managed to get an offer of $230,000 for Powder River Company's Alberta herds, the only significant remaining asset of that company. The following month he went to the annual meeting of the preferred share-

[29] Ibid, 19 and 23 Nov. 1887.

[30] Plunkett met with Streatfield and Alston on 31 Jan., he dined with Sir John Millais to discuss Geoffroy on 2 Feb., and he met with J. E. W. Booth on 27 April. Ibid., 28 and 31 Jan. 1888; 2, 4, and 6 Feb. 1888; 27 April 1888; and 19 May 1888.

[31] Ibid., 17 and 18 March 1888.

holders, who were the remaining owners of the company (the common shareholders having been wiped out). Apparently, the news regarding the Alberta herds did not improve the atmosphere of the meeting, for Plunkett's terse description was, "Usual wrangle & no result."[32]

The American Cattle Trust was also experiencing turbulence. When he got to New York the following May, Plunkett learned that the failure of the Union Cattle Company had forced Sturgis to leave the American Cattle Trust, and Lane was no longer its secretary. Dick Head was still the western manager and was unhappy that the trust was not giving him the stature he had aspired to in the cattle business.[33]

The American Cattle Trust was not a threat to competition in the cattle business, but publicity surrounding it finally did get the attention of Congress. However, when Senators George Graham Vest (Missouri) and Preston B. Plumb (Kansas) inquired about the organization, the response was that the trust was "moribund." By the spring of 1888, the internal East versus West squabbling in the trust made it impossible to chart a coherent course for the organization, and its herds only numbered about 219,000 head. When this total is viewed in comparison to the 753,000 head of cattle on the range in Wyoming Territory alone, it is obvious that the grandiose plan to monopolize the cattle industry was far from being realized.

In New York in the spring of 1888, Plunkett discussed with Dick Head the Frontier Company's contract with the cattle trust. Head told him that Frontier Company's men had delivered "many" cattle that did not belong to them, using hair brands, and Plunkett could only reply that he would look into the matter. On the way west, Plunkett stopped to see Beau Watson, who was living in Herman, where he was now keeping the accounts for Powder River Company. They spent Sunday going over the books and walking around the Herman facility. Then Horace went on to Cheyenne, where his friends at the Cheyenne Club were glad to

[32] Ibid., 27 March 1888 and 16 April 1888.

[33] John Clay said that Dick Head had "a very bad attack of swelled head" and claimed that Head aspired to be "the Napoleon of the western cattle business." Clay, *My Life on the Range*, 132. Also, *New York Times*, 5 May 1888; and Gressley, *Bankers and Cattlemen*, 264–65.

see him. There he met with Windsor, and the two men finally agreed to dissolve their partnership with Coble.[34]

Head was in Cheyenne with George W. Baxter, who was now the Wyoming manager for the trust. These two men proceeded to tell Plunkett the "inside" story of the trust, saying that Sturgis conceived the trust to save his Union Cattle Company. This explanation seemed to impress Plunkett, and after hearing their story he called the matter a "non-indictable swindle." Plunkett finally settled the dispute over Frontier Company's contract with the trust by turning over all the Frontier cattle still on the range to the trust for $15,000 cash. The sum was a considerable discount for the cattle he thought were still on the range, but he said, "Cash is a great inducement."[35]

Horace went back to the Powder River range in 1888, and when he arrived in June, he found the grass was as good as it had been in 1882 and the calf brand was "booming." Jim Winn was also on the range, and he and Plunkett rode over the mountain to Winn's old ranch, where the big house was abandoned. At the cabin of Frank Bull, Winn's former hand, Mrs. Bull gave them good food before a blazing fire, and a few days later he was reminded of the vagaries of Wyoming weather by a June snow storm. In July, Jim Winn transferred the Big Horn Cattle Company herd to the trust and then left the range for Cheyenne to look for a new job.[36]

Early in July, Plunkett went over the mountains to Paint Rock Creek, where Samuel Washington Hyatt ran a saloon that sold the "deadliest" whiskey he had encountered, and he could not resist writing down a typical cowboy story. The cowboy said he bought a bottle of the whiskey to rub on a bruise, and when he threw the

[34] Plunkett once again became partner of the WP herd when he and Windsor took a half interest in it. Plunkett received Coble's share of the cattle on the Iron Mountain ranch, and he and Windsor leased Coble's share of the Iron Mountain range. Plunkett diaries, 19–23 May 1888 and 9 and 12 June 1888.

[35] In November, the *New York Times* said, "After a short and melancholy existence this Trust died and its assets appear to have fallen into the hands of the ring member." There was little interest in the trust's certificates, which only occasionally traded on the stock exchange. Quotations for the $100 certificates were $14.625–$14.75 in the summer of 1890, compared with $16.50 a year earlier. Nevertheless, the sorry organization continued its existence for a time under different names. *New York Times*, 19 Nov. 1888 and 28 July 1890. Also, Plunkett diaries, 16 Oct.1887; 8 Nov. 1887; and 12, 13, 24, and 26 May 1888.

[36] Winn received $22.50 for the cows with calves and $28.50 for the steers, delivered to the railroad at Rock Creek. Plunkett diaries, 14–20 and 23 June 1888 and 7 July 1888.

remainder on the campfire, the whiskey promptly extinguished the fire. Although the story may have been apocryphal, it did not overstate the quality of the liquor, for the cook and some of the cowboys celebrated the Fourth of July and were afterward paralyzed.[37]

Plunkett joined the Paint Rock roundup, rising before dawn to eat breakfast. After working six hours, he consulted his watch, to learn that it was only 10:00 A.M.; by 3:00 P.M. he was tired, but the workday did not end until sunset, when most went to bed. On the days when he took a stint at night herding, he was up again at 2:00 A.M. It was no wonder that Lord Dunsany pleaded with Horace to give up roundup work. Nevertheless, he was pleased with the number of WP calves that turned up.[38]

Although Plunkett was liquidating Powder River Company, he did not give up on investing in the Powder River Basin. In July 1888, he organized the Northern Wyoming Loan and Investment Company, with George T. Beck as president, Fred G. S. Hesse as vice president, John B. Menardi as secretary, and himself as treasurer. This company was to make investments in Johnson County, and Horace at once set about writing a pamphlet to describe the new venture. The commissions Plunkett received for his work liquidating the Powder River Company provided a source of funds for the new company. Apparently the original purposes of the Northern Wyoming Loan and Investment Company were not realized, for the company was dissolved in 1891.[39]

One of the land filings made for Frontier Company in 1885 was contested in 1888, and Plunkett attended the hearing that was held in Buffalo in July. The details of the case are not known, but Plunkett said that Gilchrist, who handled the land filings for Frontier Company, had zealously gotten witnesses to swear to "all manner of

[37] Ibid., 4 July 1888.

[38] Lord Dunsany asked Horace to cable his promise to give up roundup work, and Horace sent the requested cable but followed it with a letter in which he promised to limit himself to a night or two with the roundup. Ibid., 6, 7, and 31 July 1888 and 1 Aug. 1888.

[39] The Northern Wyoming Loan & Investment Company was chartered 25 July 1888, with a nominal capital of $100,000. Plunkett's commissions funneled to the company included $2,500 paid by the Powder River Company in 1889, apparently a part of the $3^{1}/2$ percent commission on the Wibaux sale; the 1 percent commission to Courtenay had to be paid from that sum. Fred G. S. Hesse to Horace Plunkett, 22 Dec. 1889 and 6 Feb. 1891, Hesse letterbook; and Plunkett diaries, 11 and 17 July 1888 and 9 Jan. 1889.

absurdities." The great pity was that these sworn statements were unnecessary to comply with the law, but because they were in evidence, they had to be defended. The contest hearing occupied a week's time, and the atmosphere was very informal, with many irrelevant questions and answers and extensive ad libs.[40]

Plunkett's next destination was Canada, where he went to look over Lister-Kaye's new Canadian venture. By mid-August Plunkett was in Winnipeg, Manitoba, where he met Sir John Lister-Kaye and his wife to begin a fourteen-hour train ride together. Plunkett wrote his customary evaluation of Lister-Kaye, comparing him to Moreton Frewen—only more honest and "well-principled." Plunkett said that he spent the time on the trip discussing theology with Lady Kaye and talking business with Sir John.[41]

Plunkett thought Lister-Kaye was gullible and likely to be "robbed," but the fact was that Lister-Kaye had $1 million in capital in his Canadian Agricultural Coal and Colonization Company. Plunkett realized that Lister-Kaye's scheme did have promise if land values increased, but he was not impressed with the prospect for "company farming." In the fall of 1891, Plunkett again passed through the area and found the farms were doing well, although he was still convinced they would never recover the capital invested.[42]

In April 1889, Hesse cabled the news that he could sell the Powder River Company's Wyoming herds for $18.50 per head and the horses for the same price, delivered in Montana. The purchaser of the herds was Pierre Wibaux, a Montana cattleman who had fortuitously been in France (raising money to buy more cattle) when the disastrous winter of 1886–87 struck the Montana plains. Returning to Montana with the money he raised during the winter, he bought the hardy cattle that had survived the storms and made a fortune from them. Hesse reminded Plunkett that Wibaux was perhaps the only man in America who would buy such a large herd, and without waiting for contract details, Plunkett told Hesse to close the deal with Wibaux. Although the London office of

[40] Plunkett diaries, 23 and 24 July 1888.

[41] Ibid., 19 Aug. 1888.

[42] Sir John Lister-Kaye died 27 May 1924, and Lady Lister-Kaye died 13 Feb. 1943. *New York Times*, 14 Feb. 1943; and Plunkett diaries, 30 Sept. 1891. Lady Kaye was the youngest daughter of a Cuban merchant who owned Ravenswood plantation in Louisiana, and her older sister Consuelo married Lord Mandeville, who became the Duke of Manchester.

Powder River Company raised questions about the contract, other sales were being made at lower prices, and in the end the contract was approved. The company realized $139,000 from the sale of 7,419 head of cattle and 100 horses delivered in 1889.[43]

Following the sale of the cattle, Powder River Company in theory had only about 130 head of Sussex and Shorthorn cattle left and 2,240 acres of land, including 800 acres on the Tongue River. Still, the old company's cattle branded with the 76 continued to turn up on the range. In the middle of 1889, Fred Hesse noted that 170 calves had been branded, and there were "quite a lot" of steers.

In a letter written in November 1889, Fred Hesse summed up the experience of the British ranchers in northern Wyoming. "The cattle business in which I have been engaged is just about run out, it has proved another South Sea Bubble to English investors & they are all quitting the business," he said. "In the last 15 months there have been 18 Englishmen leave this country for good, so that our 'Division' is getting pretty small; of course for a few of us we cannot leave & must wait for better times."[44]

Plunkett's life also changed dramatically, for a different reason. While in Canada, he learned that his father was ill, so he cut short his trip to sail for England in early September. Lord Dunsany died in February 1889, and for the first time in years, Plunkett did not go to America that spring because of responsibilities as executor of his father's estate (he finally sailed for America at the beginning of October). The probate of Lord Dunsany's will left Plunkett financially comfortable for the first time, removing the pressure for him to make his fortune in the American West.[45]

---

[43] Pierre Wibaux, born in France in 1858, came west with the Marquis de Mores in 1883. Welsh, "Pierre Wibaux." The contract called for the delivery of all of Powder River Company's cattle (except for the purebreds) that could be located in the years 1889 and 1890, up to a total of 15,000 head, at a price of $18.50 per head. W. Courtenay, in Miles City, Mont., was apparently a broker in the transaction, entitling him to a 1 percent commission. Hesse and Plunkett also collected a commission, and Courtenay's commission was payable out of that sum. Hesse did not think the guarantee of 1,500 steers was onerous, as he expected to be able to buy any shortfall in Colorado for $15 per head. Fred G. S. Hesse, 6 June 1889, 28 Aug. 1889, and 9 Sept. 1889, Hesse letterbook. Also, Plunkett diaries, 9 Jan. 1889 and 26 April 1889.

[44] Fred G. S. Hesse to Jenny, 26 Nov. 1889, Hesse letterbook.

[45] Lord Dunsany died 21 Feb. 1889, and the will was read on 26 Feb. Plunkett sailed for New York on the *Teutonic* on 3 Oct. 1889. Plunkett diaries, 26 Aug.–5 Sept. 1888; 21 and 26 Feb. 1889; 2 Aug. 1889; and 3, 14, 17, and 18 Oct. 1889.

8

The Social Observer

When Plunkett arrived in Cheyenne in the fall of 1889, the old atmosphere personified by the Cheyenne Club was gone, and he called the city "dull & doleful," adding, "Surely the glory has departed." Notwithstanding Plunkett's depressing evaluation, the people of Cheyenne were actually enthusiastic for a different reason, as they looked forward to the end of Wyoming Territory and the birth of the State of Wyoming. Plunkett was at once drawn to this strange process, which he wanted to observe at close range. This effort marked a new direction in his association with Americans, for the inheritance from his father freed him from the necessity to seek a fortune in America. He still loved the country, and now he had the time to involve himself in matters not having to do with making money.[1]

The Wyoming constitutional convention was writing a constitution for the new state, and one of the matters to be decided was whether women should vote. In Great Britain, women could not vote in parliamentary elections, and indeed, after the Reform Bill of 1884 failed to enfranchise women, the issue came to a vote in the House of Commons only twice in the next twenty years. In the United States, although Wyoming Territory extended the vote to women in 1869, this action did not attract widespread imitation, even in the West. In fact, Utah granted the vote to women in 1870, but Congress disallowed the act in 1887. When Colorado was admitted to the Union in 1876, women's suffrage was too controversial to

[1] Plunkett diaries, 18 Oct. 1889.

be included in the initial constitution, and a referendum on the subject the following year failed. Plunkett decided to add his voice to the effort by writing a pamphlet on women's suffrage.[2]

The constitutional convention convened in Cheyenne at the beginning of September 1889 to draft the constitution for the new state. The suffrage issue was closely watched, both in Wyoming and across the country, for it was by no means certain that the new state of Wyoming would continue to give women the vote. When Melville C. Brown of Laramie (Plunkett's lawyer from the Searight case) opened the constitutional convention, he noted that the suffrage question would have to be answered but did not suggest what the answer should be. The first suggested answer was included in the draft constitution prepared by former chief justice Joseph W. Fisher, who suggested that women should continue to vote in all elections. Although the convention did not endorse the Fisher draft, George Baxter introduced an explicit proposition endorsing women's suffrage, and there was applause when this was announced.[3]

Meyer Frank, a banker from Sundance, was not one of those applauding, and he instead proposed that voters be male citizens (he later relented to permit women to vote in school elections and to hold offices related to educational advancement). Frank was a Democrat, and the *Leader,* which supported that party, predicted that his following would "not be so small as many think." However, the convention adopted Baxter's recommendation to let women vote in all elections, and the constitution that was recommended to the voters at the end of September contained this provision.[4]

Although women's suffrage was now a part of the proposed constitution, support for it was lukewarm in some quarters. The Congregational minister in Cheyenne expected the constitution to be

---

[2] Van Wingerden, *Women's Suffrage Movement in Britain*, 51–53, 55. At the end of the American Civil War, freed male slaves were given the vote, largely because of their service in the Union army. Debate on the Fourteenth Amendment largely ignored women's rights, and for that reason, many suffragettes refused to aid in the ratification of the amendment, which became effective in the summer of 1868. Marilley, *Woman Suffrage and Origins of Liberal Feminism*, 69, 71, 75. The anti-Mormon Edmunds–Tucker Act of 1887 disallowed women's suffrage in Utah. Plunkett wrote the women's suffrage pamphlet on 19 Oct. 1889. Plunkett diaries, 19 Oct. 1889 and summary for 1889.

[3] *Cheyenne Daily Leader,* 4, 6, and 8 Sept. 1889.

[4] *Cheyenne Daily Leader,* 11 and 12 Sept. 1889 and 1 Oct. 1889.

approved but criticized women's suffrage. He said, "I fail to see any great good that has been accomplished by it in Cheyenne in the sphere of morals, where we chiefly expect to see woman's power felt, though it is said to have prevented worse nominations than have [been] made." Two days later, the Roman Catholic bishop said, "I am not in favor of woman's suffrage as a principle or sentiment," but he did not ask voters to turn down the constitution.[5]

Even before he came to America, Plunkett asked Fred G. S. Hesse, up on the Powder River, to give his opinion on the matter. In June, Hesse responded that the "happiest" women "take no part in politics, except to vote as their husbands or intimate friends do." He added that women had no desire to vote, except to "help along the cause in which their male friends are the most interested."[6]

On the day Plunkett arrived in Cheyenne, the *Sun* noted that women's suffrage was the most controversial clause in the proposed constitution. Acknowledging that the convention had included the clause as a matter of right and justice, the editor said, "This opinion is probably a little in advance of the time." After defending the principle of women as voters, the *Sun* then asked any voter who intended to reject the constitution because of the suffrage clause to send the newspaper a statement of his reasons. The next day, Plunkett decided to throw some light on the subject by drafting his own questionnaire on the subject.[7]

Both of the Cheyenne newspapers printed Plunkett's eleven questions and asked readers to send their responses to Plunkett. The *Sun* editor also gave his own answers to Plunkett's questions. Following are the questions and the *Sun* answers:

> 1. About what proportion do female bear to male voters in Wyoming Territory?
> Answer: Probably one-half.
>
> 2. About what proportion of female voters avail themselves of the privilege? Can you distinguish between married and unmarried women in this respect?

[5] *Cheyenne Daily Leader*, 8 Oct. 1889; and *Cheyenne Daily Sun*, 10 Oct. 1889. The Right Reverend Maurice F. Burke, who was born in Ireland in May 1845, was appointed in 1887 as the first Catholic bishop of the Cheyenne diocese, where he faced attacks from the Know-Nothings Party and even suggested that the diocese be closed.

[6] Fred G. S. Hesse to Horace Plunkett, 30 June 1889, Hesse letterbook.

[7] *Cheyenne Daily Sun*, 18 Oct. 1889; and Plunkett diaries, 19 Oct. 1889.

Answer: Three-fourths. Should say that married women vote more generally than unmarried.

3. Do women vote independently, or do they follow their husbands or other male relations with whom they live?
Answer: Largely influenced by their male relatives, but would resent dictation.

4. Are women more influenced by personal, as distinct from political considerations, than men in the selection of candidates for office?
Answer: They are. The character of the candidate cuts quite a figure with women. Personal solicitation has also an appreciable effect.

5. What is their influence as regards political corruption and generally what moral effect (if any) have they on the tone of politics?
Answer: The "women vote" is usually regarded as more readily obtainable by candidates possessing such virtues as integrity sobriety and decency and this fact is not disregarded in the nomination by political parties.

6. Do women appear anxious to press the claims of their sex to a larger share of official employment?
Answer: They do not.

7. Do women take an interest in (1) local, (2) territorial, (3) federal politics which can be traced to female suffrage?
Answer: Their interest is in the order indicated.

8. Do women attach any great importance to the suffrage, and would they generally resent any proposal for its repeal?
Answer: Apparently they do not. Yet they would quite generally resent any proposal to repeal the law.

9. What influence (if any) has the political equality of the sexes exercised on family ties and on the view with which the marriage life is regarded?
Answer: It cannot be said to have had any injurious effect. Our opinion is that women gain the respect of men in proportion as they gain power, usefulness and intelligence, and just so far as the ballot contributes to these is a benefit to women—and indirectly to man.

10. On what issues and on what occasions have women affected, directly or indirectly, the course of Wyoming legislation?
Answer: There has been no issue before the people in which women have taken a special interest.

> 11. Can you give any further information tending to show the advantages or disadvantages of female suffrage to the territory of Wyoming? Answer: It gives greater power to home-makers, which is the conservative element. It is this element which is most apt to oppose public extravagance and excesses and extremes of all kinds. After twenty years' experience I cannot say that the effect of woman suffrage is very marked. It has not affected, perceptibly, the marriage ties, social relations nor the political status of any portion of the territory. Women are conceded suffrage because it is regarded as their right and not for any special use they may make of it.

A week after his questionnaire appeared in the newspapers, Plunkett noted that the people of Wyoming were puzzled by it.[8]

Plunkett pursued his research by attending a "Literary Circle" at M. C. Brown's house to get information from a large group of women, and he conducted a number of interviews with Cheyenne residents. On the eve of the territorial election, where the voters approved the new constitution, Plunkett wrote a letter to the *Leader,* urging approval of the constitution. This letter is a good example of the Plunkett style of advocacy, offering something to both sides in a controversy and thus generally leaving both sides unhappy with him.

Plunkett began his letter by noting that he had at first opposed statehood, because he thought it was "an officeholder's scheme and a boomer's snare." However, after reading the constitution he had changed his mind, saying that it established a "purer" form of government. Then he went on to discuss that "great stumbling block," women's suffrage. He would have preferred that the issue be submitted to the voters as a separate referendum, but now that it was imbedded in the constitution, he did not think any politician would dare to propose that it be removed. "Therefore," he said, "we shall have this feature of our polity whether we wish it or no." It was hardly a ringing endorsement of the woman voter.[9]

The issue of women's suffrage continued to embroil politics in the United States into the next century. In the Wyoming election that approved the new constitution, over 700 women in Cheyenne

---

[8] *Cheyenne Daily Leader,* 23 Oct. 1889; and *Cheyenne Daily Sun,* 23 Oct. 1889. Also, Plunkett diaries, 26 Oct. 1889.

[9] *Cheyenne Daily Leader,* 5 Nov. 1889.

cast ballots. The women's suffrage clause also survived the congressional debates on the admission of Wyoming (although the votes were close on that issue), and the State of Wyoming was admitted in the summer of 1890. Although the women's suffrage issue had been decided in Wyoming, it was still a hot issue on both sides of the ocean, and Plunkett took notes on the 1889 election in preparation for his article on women's suffrage.[10]

Plunkett's article on women's suffrage, "The Working of Woman Suffrage in Wyoming," appeared in the May 1890 issue of the influential journal *Fortnightly*. He began his article by tracing the history of the 1869 Wyoming territorial law that gave the vote to women, basing his comments on interviews with a number of men who had served in the 1869 legislature. Although he did not get a chance to speak with Col. William H. Bright, who introduced the 1869 bill, Plunkett declared that none of the survivors of that first session whom he had interviewed could recall any discussion of the general principle of women's suffrage. Instead, these legislators commonly described the suffrage vote as a joke that would be good advertising for the territory, which badly needed immigration.[11]

Indeed, Plunkett noted that few of the men in the territory in 1869 were married, and that the women were "a pretty hard formation" (p. 5) of the social strata. Nevertheless, although the decision to grant women the vote was taken in haste and without great deliberation, none of the residents of Wyoming seemed to regret their action, and Plunkett correctly predicted that future historians would see it as the "triumph of a great principle" (p. 6).

One of the early results of the 1869 law was the appointment of women to juries, although ironically, the law itself did not in fact

---

[10] Plunkett diaries, 5 Nov. 1889. Wyoming was admitted to the Union on 10 July 1890. The constitution was approved by 76.5 percent of the 8,195 votes cast, carrying nine of the ten counties then in existence (Sheridan County voted against the constitution). By 1900, universal suffrage existed only in Colorado, Utah, Idaho, and Wyoming; referenda had failed in California, Washington, South Dakota, and Oregon. Marshall, *Splintered Sisterhood*, 152. An amendment by Rep. William McKendree Springer (D-Ill.) to give Wyoming voters an explicit option to reject women's suffrage failed by only six votes in the House, and a second amendment to strike the women's suffrage provision in the Wyoming constitution failed by the same margin.

[11] Plunkett diaries, 5 and 6 Nov. 1889. Although Plunkett repeated the story that a woman had influenced Bright in favor of the bill, he did not mention Mrs. Esther Morris's name in the article.

extend that privilege to women. Plunkett noted that a mixed grand jury had insisted on enforcement of the Sunday closing law for saloons, which he took to be a sign that women had a higher sense of moral judgment than the men they replaced on the jury. He also asserted that women were not so easily bribed as men.[12]

Some had argued that female jurors would prove to be too gentle or susceptible to deal with criminal trials, but Plunkett found this was not so. A mixed jury in a murder trial issued a guilty verdict against a handsome defendant and sentenced him to ten years even though his demeanor was pale, dreamy, and Byronic (to use the description of a female juror). Indeed, a lawyer familiar with the case declared that only the presence of *men* on the panel prevented the "bloodthirsty" (p. 7) women from making the sentence even more severe.

On the basis of his unscientific survey, Plunkett concluded that qualified women voters accounted for about one-third of the electorate, and that 80 percent of the eligible women did vote. He also found that more married women voted than did unmarried women, and he ventured the rather quaint suggestion that this factor was influenced by a reluctance of spinsters to declare they had reached their majority, at least until after they had reached middle age.

Another finding of Plunkett's survey was that female voters did not extend any preference to female candidates, although he still believed that the physical appearance of a candidate was more important to women voters than to men. Moreover, women would not vote for a candidate with a "stain" (p. 13) on his record. Finally, these considerations made it much more likely that women would split the ticket, rather than vote the party line.

On the subject of other issues of equality between the sexes, Plunkett noted that women did not use the franchise to press for a larger share of employment. He attributed that factor to their undoubted equal qualifications *"for the work they undertake"* (p. 13, emphasis added), which he defined as teaching and clerical occupations.

---

[12] One of the peculiarities of the brief 1869 law was the language that gave women the rights of electors, insofar as voting and holding office were concerned; a separate enactment required that juries consist of males, and this law was not amended. After women had served in three terms of court, it was pointed out that this service was illegal, and it was discontinued.

Moreover, he grandly declared that women had no need to take action to protect or extend their legal rights, "the feeling of American men requiring no pressure to induce them to place women in a position of equality with themselves" (p. 14).

In summary, it was Plunkett's thesis that women's suffrage in Wyoming could be characterized at worst as the aftermath of a joke and at best as a pragmatic advertising gimmick, which nevertheless conferred a "political windfall" (p. 14) on the female voter. Without these anomalous factors, Plunkett doubted that the concept of women's suffrage could have overcome the apathy of the women themselves.

Speaking to those (such as the British suffragettes) who might want to use the Wyoming example in their own regions, Plunkett warned that this would be unwise, "until a state of things so wholly exceptional has passed away" (p. 17). As for the future of the female franchise in Wyoming, he thought the only danger to it was "the possible attempt . . . to force the women of Wyoming to play a part which would be so distasteful to them as to make the privilege become a burden" (p. 2).

Plunkett had the *Fortnightly* article reprinted in Cheyenne and apparently distributed it fairly widely in that region. Although Horace undoubtedly believed that he had written a balanced analysis on the matter of extending the franchise to women, he was soon to learn that he had ignited a controversy with the women's suffrage movement, and a battle of the pamphlets ensued.[13]

Hamilton Willcox, who was an official in the New York Woman Suffrage Party, published a pamphlet in November to respond to and correct Plunkett's work and to supply "omissions" in other works on women's suffrage. Willcox objected to Plunkett's assertion that Americans were "considerate to women." Willcox declared that a "considerate" population would not have imprisoned 20,000 women in New York the previous year or created nearly 100,000 illegitimate mothers each year.[14]

---

[13] Plunkett, *Working of Woman Suffrage in Wyoming*.

[14] James K. Hamilton Willcox was a New York lawyer, and the fulsome title of his pamphlet was *Wyoming: The True Cause and Splendid Fruits of Woman Suffrage There, from Official Records and Personal Knowledge; Correcting the Errors of Horace Plunkett and Professor Bryce, and Supplying Omissions in the History of Mrs. Stanton, Mrs. Gage and Miss Anthony, and in the History of Wyoming by Hubert Howe Bancroft; with Other Information About that State.*

Next, Willcox challenged Plunkett's assertion that American men did not have to be pressed to accord women a position of equality. Willcox fumed that the common law disabilities of women were enforced in every state, and American men resisted removing them, so that if a woman were on trial she could not be judged by a jury that included women. He also made the interesting point that when a male criminal went to jail as punishment for his crime, his innocent wife was also punished by being deprived of his support.

Willcox also criticized Plunkett's history of the suffrage bill in Wyoming, which Plunkett had claimed was in part a joke and amounted to a "political windfall" for women. Willcox claimed that the Wyoming enactment was a direct descendant of the women's suffrage agitation that began with a bill introduced in Congress at the end of 1868. Aside from this assertion, Willcox contributed little of substance to Plunkett's description of the passage of the act.[15]

Willcox criticized Plunkett's characterization of women's suffrage as an "experiment" and "wholly exceptional" and pointed out that women could vote for all elective offices in Britain—except for those in Parliament—leaving the impression that women's suffrage was favored in the United Kingdom. This was a considerable exaggeration, for British prime minister Gladstone opposed granting the vote to women, saying it would "trespass" on the delicacy, purity, and refinement of women, and the Queen, herself, also opposed women's suffrage. The Prince of Wales was of the same mind as the Queen on this issue, and he considered the behavior of the British suffragettes "outrageous."[16]

Plunkett continued to research the subject of women's suffrage, and when he was in Cheyenne in the fall of 1890, he met with Theresa Jenkins, a leader in the movement. Mrs. Jenkins, a life-

[15] HR 1531, to extend the franchise to women in the territories, was introduced on 14 Dec. 1868 by Congressman George Washington Julian of Indiana and was immediately referred to the Committee on Territories, where it died.

[16] The Queen also did not want women working in the dissecting rooms, studying things that "cld not be named before them." Longford, *Queen Victoria: Born to Succeed*, 395. After he had ascended the throne, King Edward commented to Daisy Fingall, "You can get all you want without the vote." Leslie, *Marlborough House Set*, 303. Also, Magnus, *King Edward the Seventh*, 390.

long friend of American suffragette Carrie Chapman Catt, cast her first vote in the Wyoming territorial election of 1869.[17]

Although he was busy with the women's suffrage issue in late 1889 and in 1890, Plunkett also attended to other matters in America. At the end of October 1889, he was in Cheyenne, and Gilchrist hosted a dinner that continued until 3:00 A.M. Jim Winn, Ralph Stuart-Wortley, and Fred Hesse came to Cheyenne to discuss business, but they arrived late, causing Horace to grumble that he didn't have time to go to Washington, D.C., "& other places" as he had planned. Jim Winn had a drinking problem and so was not any help with the discussions, but two days later he hosted a dinner for Horace, who said Winn was an excellent host.[18]

After finishing his business discussions in Cheyenne, Plunkett left for Cleveland, Ohio, where he and Ralph Hickox discussed investments in West Virginia coal and iron lands. In New York, Horace had a long meeting regarding the American Cattle Trust, which was being reorganized as a corporation. The fortunes of the trust continued to worsen, although the directors did declare a 3 percent dividend on 1889 operations, despite the fact the herd count had declined by more than 50,000 head. Packing operations showed a loss, as the trust operations were not competitive with Swift and Armour, and the packing plant was returned to Morris at a substantial loss to the trust. Moreover, the feeding operation at Gilmore also had a huge loss.[19]

In the summer of 1890, the Western Union Beef Company was organized in Denver to take over seven ranches formerly owned by the trust in Wyoming, Colorado, Texas, and New Mexico. Charles M. McGhee, who was George Baxter's father-in-law, was president of the new corporation, and both men were directors of

[17] Plunkett diaries, 26 Oct. 1890.

[18] Plunkett diaries, 31 Oct. 1889 and 2 Nov. 1889.

[19] John Alexander Osborne, who worked for the Western Union Company in this period, estimated that the company was running 300,000 head of cattle. This total must have been inflated by a large number of cattle that were being managed for others. In 1890 Baxter was running the Bar C Ranch of Peters & Alston, paying the taxes on the cattle as rental on the property. Fred Hesse wrote to Baxter in the spring of 1891 to ask if the latter wanted to take the Bar C again in 1891 on the same basis. The trust carried only 164,000 head on its books in 1889, and even that total may have been overstated. Osborne, "John Alexander Osborne," 316; and Fred G. S. Hesse to George W. Baxter, 15 April 1891, Hesse letterbook.

the company. Warren had resigned as Wyoming manager of the trust and was replaced by George Baxter. Baxter was a Democrat, and there was bad blood between him and Warren, a Republican. In the summer of 1890, when Baxter was running against Warren, a letter in the Cheyenne *Sun* blasted Baxter's "fat" salary and accused him and his relatives of controlling the trust. An accompanying editorial vouched for the truth of the anonymous letter and asserted that Baxter's father-in-law, Charles M. McGhee, was buying up the trust certificates at 2 1/2¢ on the dollar, or 10 percent of the original issue value.[20]

The $100 shares of Western Union Beef traded in New York but were apparently never listed on the exchange; from time to time, bid and asked prices for the shares ranged from $4 1/2 to $9 per share, and the number of shares traded was minor. Obviously, the market would not give this security even the 25 percent value that was the basis for acquisition of herds by the American Cattle Trust.[21]

Plunkett sailed for home in mid-November 1889, and at the end of the month he set out on a trip to Europe and Africa. He was back in London at the end of December to meet with Frank Kemp to talk over Omaha real estate investments. A sure sign of his improved financial condition after his father's death was the fact that early in 1890 he was able to buy out Willie Blacker's investments in America. The remnants of the Wyoming ranching operations continued to take up time, and in March, Jim Winn came by to discuss winding up the remnants of the WP and Bar X Bar herds in Wyoming.[22]

Plunkett returned to America at the beginning of October 1890, and in New York he met Ralph Stuart-Wortley, who came up from Portsmouth, Virginia, where he was the vice president of a

[20] The Western Union Beef Co. was organized in Colo. with an authorized capitalization of $15 million, consisting of 150,000 shares. The board members were John L. Routt, George W. Baxter, Richard T. Wilson, Charles M. McGhee, G. G. Symes, Samuel T. Thomas, M. O. Wilson, and John G. Moore. Plunkett diaries, 10–13 and 30 Nov. 1889 and 1–28 Dec. 1889; Gressley, *Bankers and Cattlemen*, 259ff; and *Cheyenne Daily Sun*, 24 Aug. 1890.

[21] Late in 1892, Plunkett learned that the $100 shares of the Western Union Beef Co. were quoted at $5. Plunkett diaries, 10 Nov. 1892. There are a few quotations for Western Union Beef as an unlisted stock on the New York exchange from 1893 to 1896.

[22] Plunkett diaries, 27 Jan. 1890 and 27 March 1890.

small railroad. Then, on the way west, Horace stopped for two days at Cleveland to see Ralph Hickox, and on his return in November, the two men agreed to risk $10,000 on a joint speculation in an Illinois steel company. In Omaha, Plunkett saw Windsor and Frank Kemp, and Windsor proposed a scheme to generate electricity by water power to supply Omaha, but Plunkett did not like the idea. In Cheyenne, Plunkett met Boughton, Chaplin, and Gilchrist before going to a meeting of the Wyoming Development Company.[23]

On the way east on the train from Cheyenne to Omaha, another social issue caught Plunkett's attention. The Prohibition Party, which had fielded a candidate for governor of Nebraska in every election since 1884, once more had the issue before the voters in a campaign that produced more excitement than the state had ever seen. Plunkett sounded out people on the train regarding prohibition and found the farmers generally favored it. On election day, he pronounced American machine politics "unedifying," and thought the "ruffianism" at the polls was disgraceful. Despite all the fuss, the issue on prohibition was never in doubt, as the voters gave a 23,000 margin to the wets.[24]

On November 11, Plunkett sailed for England. When he next returned to America in the fall of 1891, he came on a mission for the British government. Arthur Balfour, chief secretary for Ireland, appointed Plunkett to the Congested Districts Board, which was concerned with the welfare of the Irish peasants in the so-called congested districts, mostly in the west of the island. One of the proposed solutions to the problem was to promote emigration to the United States and Canada, and Plunkett was sent over on a tour of inspection.[25]

He brought with him his distant cousin, Lord Fingall, and they first traveled to Ottawa and then to Omaha, where they talked with prominent Irishmen about immigration. While they were in

---

[23] Plunkett diaries, 9–19 Oct. 1890. I assume Ralph Stuart-Wortley was vice president of the Seaboard Air Line Railway, which had its headquarters at Portsmouth. The *New York Times* of 12 Aug. 1888 said this little-known railroad was "rich."

[24] Olson, *History of Nebraska,* 224, 226; and Plunkett diaries, 29 Oct. 1890 and 4 and 5 Nov. 1890.

[25] At the end of the year, the Scottish government appointed Plunkett to its Colonisation Board. Plunkett diaries, 31 July 1891, 23 Sept. 1891, 11 and 19 Oct. 1891, and 31 Dec. 1891.

Omaha, Plunkett and Fingall witnessed a lynch mob, giving Plunkett the chance to study one of the more savage aspects of American life. The media played no small part in fanning the public anger, even before Joseph Coe, a black man, was apprehended for raping Lizzie Yeates, a girl of five. The *Omaha World Herald* called the matter "a well matured subject for lynching" when the false rumor later circulated that the little girl had died, and public indignation was much inflamed. The *Omaha Excelsior* later claimed that one of the dailies wrote a "complete" report of the lynching before it had occurred.[26]

There was on-the-spot press coverage of the mob that surrounded the jail, a crowd of 10,000 or 15,000, including the mayor, chief of police, and former governor James E. Boyd. Although press reports claimed that the mob overwhelmed the police and firemen, the *Omaha Excelsior* noted that the crowd was not unruly, and when one officer threatened to shoot the rioters, the crowd backed off. Fingall and Plunkett watched the horrible spectacle, which Horace called "this strange institution of a great people," as the mob dragged Coe from the jail, beat him, and hanged him from the trolley wire.

Plunkett declared that he never wanted to see such a scene again, "but it was worthwhile to see it once." Then, in typical fashion, he wrote his analysis of the incident. "It was partly indignation, more cruelty," he said, then added, "The institution [i.e., lynching] must prevail until the administration of the law is true & *sure*." Although the Omaha press later pronounced the lynching a stain on the city, the *Omaha Bee* noted that it was a typically "Nebraska" action.[27]

Plunkett was called home early and had to compress his personal business into two days at Omaha, sailing from New York on the *Teutonic* on October 21, 1891. In the year before he again returned to the United States, his old ranch territory was the scene of a shocking armed invasion intended to deal with cattle rustling. Cattle rustling was a problem that had been brewing in northern Wyoming for a number of years.

---

[26] *Omaha Excelsior,* 17 Oct. 1891.

[27] Plunkett diaries, 9 Oct. 1891; *Omaha Excelsior,* 17 Oct. 1891; and *Omaha Daily Bee,* 10 Oct. 1891. A Presbyterian clergyman in Kansas City said the mob should have let the law take its course but also praised Omaha for "an arousing of the public morals."

In 1889, Fred Hesse, who had been foreman for Frewen and was still winding up the Powder River Company's affairs for Plunkett, became involved in a group effort to stem the worsening cattle rustling problem in the Powder River basin. Hesse commented to Beau Watson, "This is getting to be a hard country to live in." When the cattlemen offered a $1,500 reward for horse and cattle thieves, Hesse—without even consulting with Plunkett—agreed to participate in that expenditure. Each ranch's proportion of the cost of the reward was based on its number of cattle, and later, the Powder River Company's share had to be increased when Colonel Pratt withdrew from the pool.[28]

Hesse also engaged in his own detective work, paying Si Wishert $50 a month to infiltrate the rustler gang and supply information. George Baxter, apparently acting on behalf of the Western Union Beef Company, employed Wishert to do the same sort of work, and the Livestock Commission was certainly aware of it. Wishert spent the summer of 1889 "trying to get in with the gang," and by November, Hesse concluded that Wishert had gained their confidence. There can be no doubt that both Moreton Frewen and Horace Plunkett shared Hesse's views on the cattle theft problem, so it is not necessary to inquire whether Hesse acted on his own or in his employers' interests.[29]

Because of his activism, Hesse became a focal point of opposition by the so-called "rustler" group. At the end of March 1892, the manager of Hesse's 28 Ranch found a pile of the hides and bones of twelve cows and steers, most of which had been shot in the head. Nearby, a 76 cow and a horse had been also shot. Clearly, someone was sending Hesse a message.[30]

In April 1892, an expedition of cattlemen and hired Texas gunmen invaded Johnson County to put down cattle rustling there, and Fred Hesse met the special train carrying the invaders from Cheyenne. Acting on a tip, the invasion party went to the KC

---

[28] Fred G. S. Hesse to Horace Plunkett, 22 Dec. 1889 and 8 Feb. 1890, Hesse letterbook.

[29] Hesse emphasized that he did not want Wishert's name to appear in any public record, and for this reason, he did not want to give the man any other duties. Fred G. S. Hesse to R. F. Glover, 15 and 24 Nov. 1889, Hesse letterbook. Both Hesse and Glover were livestock commissioners for Wyoming Territory.

[30] E. C. Miller to Fred G. S. Hesse, 27 March 1892, and Sheriff J. M. Ramsey to Fred G. S. Hesse, 28 March 1892, Hesse letterbook. At the time of the invasion of Johnson County, Fred Hesse was state livestock commissioner in Johnson County.

ranch, one of the Powder River Company's line camps, expecting to find fourteen or fifteen rustlers there. After a night ride through a fierce storm, they arrived to find only two wanted men and two cowhands who were in the wrong place at the wrong time.

One of the men in the KC cabin was Nate Champion. Nathan David Champion apparently came north with a trail herd in 1881 and was hired at Plunkett's EK ranch. Later, he was wagon boss at Peters and Alston's Bar C, but by 1886 he was back at the EK, where Plunkett laid him off for lack of work. Nevertheless, he was again working at the EK in 1888, when he was apparently fired by Baxter, who was then acting for the cattle trust. After that, Nate was allied with the four men who owned the Hat brand, a group regarded as the most notorious rustlers in Johnson County. Near the end of 1891, four men tried to kill Nate, who was then living in a cabin not far from the old Bar C headquarters. Nate survived that incident and moved to the KC cabin, where the invaders found him. After a siege, which Nate described in a pocket notebook, both Nate Champion and Nick Ray were killed. The invaders were subsequently pinned down by local settlers at the TA Ranch and finally were rescued by the U.S. Army and incarcerated at Fort D. A. Russell, near Cheyenne.[31]

Hesse was with the invaders who were held at Fort D. A. Russell, and soon he was in contact with the outside world. Operating from Omaha, Frank Kemp took care of necessary details for Hesse, locating his valise and overcoat, opening the mail (being diverted to Omaha), and sending money to Mrs. Hesse, who took the children to North Platte, Nebraska, where she had relatives.[32]

Hesse had no difficulty carrying on business from the fort, and indeed, his correspondents sometimes asked him to check business matters with his fellow prisoners. Hesse was soon busy writing letters, and by May, Kemp was able to make tentative arrangements to safeguard the remaining assets in northern Wyoming, including

[31] The Hat brand was owned by Jack Flagg, Martin Allison Tisdale (alias Al Allison), Billy Hill, Lewis A. Webb, and Thomas Gardner. Smith said all of these men came from Williamson County, Tex, which was Nate Champion's birthplace. Nate was not the only Champion family casualty in this affair. His younger brother, Rufus Dudley Champion, was killed by Mike Shonsey on 23 May 1893. Helena Huntington Smith, *War on the Powder River*, 114, 154–55, 282.

[32] Frank A. Kemp to Fred G. S. Hesse, 22 April 1892, Hesse letterbook. Peirce may well have gone to Omaha for safety's sake, as he received a telegram there on April 29.

arrangements for the spring roundup. Early in August, Kemp visited Cheyenne and saw Hesse, and in November he was writing to Hesse at North Platte, where Mrs. Hesse and the children had been living since the Invasion.[33]

We know that the cost of defending the invaders was heavy, and early in 1893, Frank Kemp notified Hesse that he had received a request from Charlie Campbell for an additional $500 "towards expenses." Because Campbell was working for John Clay, Jr., president of the Wyoming Stock Growers Association, presumably these expenses were for the defense of the cattlemen. After the cases against the invaders were dismissed and Hesse was released, he did not feel safe returning to the Powder River Basin and instead moved with his family to North Platte, Nebraska. When Plunkett came to Omaha near the end of 1892, Hesse came there to report first-hand on the events of the past year.[34]

After the invasion, Plunkett received accounts of the affair from his old friends, some of them in detail, but none have survived to throw additional light on what is still a murky record. Plunkett did not commit to paper any of the things he learned, beyond noting that his "old friends," Hesse, Teschemacher, Irvine, Canton, and de Billier were involved. As to the culpability of the intended targets for the raid, Plunkett had no doubt, for he said that the rustlers "have taken hold of the whole Johnson County & stolen most of the cattle."[35]

---

[33] Frank A. Kemp to Fred G. S. Hesse, 7 and 14 May 1892; W. P. Keays to Fred G. S. Hesse, 10 May 1892; and Frank A. Kemp to Fred G. S. Hesse, 9 Aug. 1892 and 7 Nov. 1892, Hesse letterbook.

[34] Frank A. Kemp to Fred G. S. Hesse, 20 Jan. 1893; and Charles H. Burritt to Fred G. S. Hesse, 23 Jan. 1893, Hesse letterbook. Hesse wrote a long paper on the events leading up to the invasion for the use of Maj. Frank Wolcott, who had apparently requested it. There are several drafts of Hesse's paper, both in the Fred Hesse collection in Buffalo and in the Fred G. S. Hesse collection in the American Heritage Center. Frank Kemp to Fred G. S. Hesse, 14 and 20 May 1892 and 8 July 1892, Hesse letterbook. Also, Plunkett diaries, 21 Nov. 1892.

[35] Hesse wrote to Horace late in April 1892, and later in the year when Plunkett was in Omaha, both Hesse and Teschemacher saw him and told him about the raid. Hesse was then living "in exile" in North Platte, Neb. Plunkett diaries, 29 and 30 April 1892 and 15 and 21 Nov. 1892.

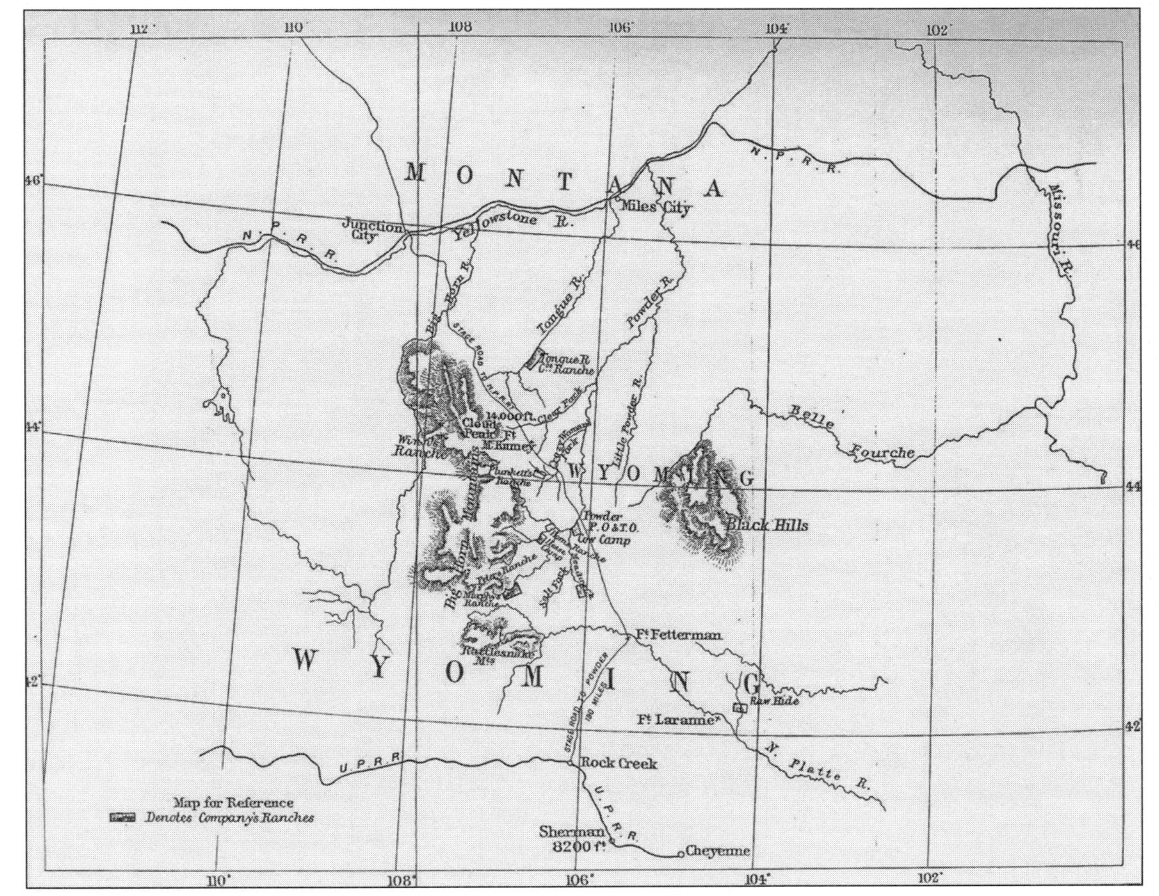

The Powder River Basin This map, prepared for Moreton Frewen, shows the location of British ranches in the Powder River Basin in the 1880s. "Winn's Ranche," west of the Big Horn Mountains, is that of the Big Horn Cattle Co. Plunkett's ranch is shown directly south of Ft. McKinney; farther south is "Peters Ranche," the ranch of Peters and Alston; and the several ranches of Frewen's Powder River Cattle Co. are also shown. *Author's collection.*

Horace Curzon Plunkett as a young man.
Courtesy Lady Dunsany.

Horace Plunkett on horseback on the Wyoming range.
Courtesy Lady Dunsany.

Sir Horace Plunkett in March 1923.
Courtesy American Heritage Center, University of Wyoming, Horace Plunkett photo file.

Moreton Frewen, April 1878. This photograph was taken in New York when Frewen was on his trip to the West that inspired his venture in Wyoming cattle ranching. *Author's collection.*

(*seated, holding hat*) Fred George Samuel Hesse, foreman of the Powder River Cattle Co, Ltd; (*standing behind him, leaning on chair*) Moreton Frewen. *Courtesy American Heritage Center, University of Wyoming, Fred Hesse papers.*

Ralph Granville Montagu-Stuart-Wortley (younger brother of the 2nd Earl of Wharncliffe), who ranched in the Powder River Basin in the 1880s and later moved East, where he was a stockbroker at the time of his death. *Courtesy Richard Alan Montagu Stuart Wortley of Cumberland, Maine, 5th Earl of Wharncliffe and great grandson of Ralph Stuart-Wortley.*

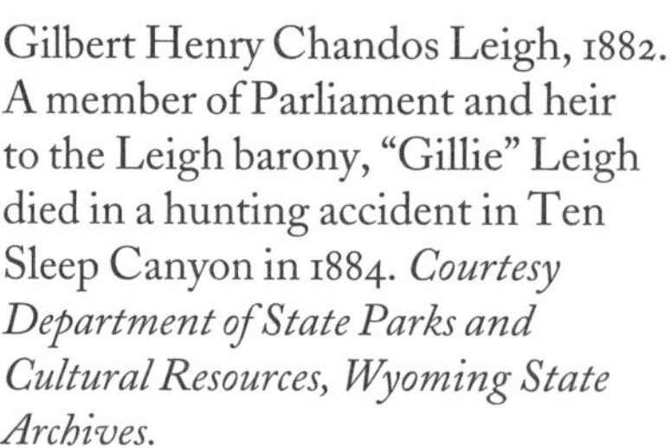

Gilbert Henry Chandos Leigh, 1882. A member of Parliament and heir to the Leigh barony, "Gillie" Leigh died in a hunting accident in Ten Sleep Canyon in 1884. *Courtesy Department of State Parks and Cultural Resources, Wyoming State Archives.*

LEIGH CREEK MONUMENT

This memorial to Leigh was erected by Horace Plunkett atop the cliff in Ten Sleep Canyon where Leigh plunged to his death. The inscription reads, "In Memoriam, Gilbert Leigh, Sept. 14, 1884. Erected by Ranchmen." Leigh Creek, which flows into Ten Sleep Creek at this point, is named for Leigh. The sign nearby on the highway erroneously refers to Leigh as a "remittance man." *Author's collection.*

Henry Fitz-Gerald Heard, ca. 1960, Plunkett's onetime companion and his executor. *Courtesy Jay Michael Barrie, © The Barrie Family Trust.*

Plunkett's friend Arthur James Balfour, first lord of the treasury until 1902; prime minister, 1902–1905; first lord of the admiralty in 1915, and foreign secretary, 1916–19. He was created Earl Balfour in 1922. *Courtesy the National Portrait Gallery, London.*

Plunkett's friend in the agricultural cooperation movement in the United States Dr. Charles R. McCarthy, head of the Wisconsin Legislative Reference Library, Madison, Wisc., with his staff. *Courtesy Wisconsin Historical Society, Image No. WHi-3783.*

Plunkett's associate in the Wyoming Development Company's Wheatland Project Joseph Maull Carey, mayor of Cheyenne, 1881–85; Wyoming territorial delegate to Congress, 1885–90; Wyoming senator, 1890–95; and Wyoming governor, 1911–15. *Courtesy Department of State Parks and Cultural Resources, Wyoming State Archives.*

Robert Davis Carey, son of Joseph Maull Carey, also worked with Plunkett in the Wheatland Project. He was Wyoming governor, 1919–23, and senator from Wyoming, 1930–37. *Courtesy Department of State Parks and Cultural Resources, Wyoming State Archives.*

Plunkett's friend Rt. Rev. James B. O'Connor, first bishop of the Roman Catholic Diocese of Omaha. Their conversations were mostly about real estate, not religion. *Courtesy Douglas County Historical Society Collections, Omaha, Neb.*

Col. Edward Mandell House, Woodrow Wilson's trusted advisor, who was Plunkett's contact in the Wilson administration. *Courtesy the Center for American History, University of Texas at Austin, PPC file.*

9

New Directions

PLUNKETT RETURNED to America in 1892, and once again an important reason for the trip had nothing to do with his own business interests. As a member of the Irish Industries Association, he was assisting with preparations for the Irish exhibit at the Columbian Exposition, which was to open in Chicago in 1893. On the way over he talked with a fellow passenger who was also interested in the coming exposition. William Broderick Cloete, a wealthy Oxford classmate of Plunkett's brother, raced yachts and had extensive real estate holdings in northern Mexico and on the side was financing the work of Hiram Maxim. Maxim was the American-born inventor of the machine gun (which had been adopted by the British army the year before), and his attention was now focused on a flying machine he had invented.

Cloete told Plunkett that he had suggested that Maxim should turn his inventive genius to solving the problem of aerial navigation, and Maxim replied, "If a goose can fly, surely a man can." Maxim began building the machine in 1891, and at the time of Cloete's conversation with Plunkett, he was hoping to have it ready to exhibit in Chicago the following year. Although Maxim did not make his hoped-for deadline, in 1894 he did succeed in getting his machine to lift off the ground for a short distance.[1]

---

[1] Plunkett diaries, 4 Nov. 1892. Cloete was thirty-eight at the time of his Atlantic crossing with Plunkett. In the fall of 1885, he and Robert R. Symon of the Mexican Central Railway purchased the Sanchez grant of 1.4 million acres near Monclova in Coahuila state. Cloete was one of those who died when the *Lusitania* was sunk by a German submarine on May 7, 1915.

The Columbian Exposition, to celebrate the 400th anniversary of Columbus's voyage to the New World, was awarded to Chicago by Congress in 1890, and the organizers of the fair selected a 645-acre site near Lake Michigan, making it the largest world's fair up to that point. The 1876 Centennial Exposition in Philadelphia, the first large fair in the United States, attracted ten million people, and the world's fairs in Paris in 1878 and 1889 were also a great success. The organizers of the Chicago exposition expected to surpass those earlier numbers, and their original plans called for 15 million visitors. In fact, even this plan proved too modest, for 27.5 million attended, and more than 21 million of that total paid the 50¢ admission.

The Irish Industries Exhibit was a project of the Irish Industries Association, founded by Lady Aberdeen while her husband was Viceroy of Ireland in 1886. Plunkett was a member of the association and did much of the spadework on the exhibit; he found Lady Aberdeen difficult to work with. In 1890, the Aberdeens visited Chicago to lay the groundwork for the construction of the Irish exhibit at the fair, and a Chicago committee was formed. The *Chicago Herald* opened a subscription list to take contributions to support the exhibit, and Andrew Carnegie and Marshall Field both contributed.[2]

Eventually, there were two Irish exhibits. The two were located on either side of the Illinois Central railroad tracks, some distance from the British exhibit, which was over by Lake Michigan. The 25¢ admission entitled visitors to receive a piece of Irish sod, to taste authentic Irish stew, and to kiss the Blarney Stone at Blarney Castle, which had been constructed on an exact two-thirds scale. The entrance to the Irish Industrial village was through a reproduction of the St. Lawrence Gate at Drogheda, which was built in 1200. A copy of the Donaghmore Cross was in the village square, and those who climbed the fifty-foot tower of the castle also had a spectacular view of the main fairgrounds to the east.[3]

---

[2] Ishbel, *We Twa*, 299–300, 309; and Gordon, *Wild Flight of Gordons*, 189. Lady Aberdeen was the daughter of Lord Tweedmouth, who was the largest shareholder of the Rocking Chair Ranch in Texas and owner of the Horse Shoe Ranche in Dakota Territory. When Plunkett met Lady Aberdeen in the fall of 1891, he wrote, "A good woman, I think, but rather scattering. Still she is rich & attracts wealth into the sphere of usefulness." Plunkett diaries, 26 Aug. 1891; and Woods, *British Gentlemen in the Wild West*, 111.

[3] Bolotin and Laing, *World's Columbian Exposition.* The village also featured a reproduction of Donegal Castle, the ancient seat of the O'Donnells, the earls of Tyrconnell, and there was a reproduction of the Antrim Round Tower.

Judging from Lady Aberdeen's description of her work in organizing the project in Chicago, one might suppose that the financing of the project had been arranged, but this was not so. In the fall of 1892, Plunkett stopped in Chicago to negotiate two concessions for the Irish exhibit, and he learned that there were no funds available to pay the required concession fee. Horace and five others were forced to subscribe £5,500 on the spot, "pour encourager les autres" (to encourage others).[4]

The Chicago exposition opened in 1893, and in September Plunkett came over again to "investigate" the Irish participation in the fair. He was at the fair for Chicago Day, where the "perfectly orderly" crowd of some 700,000 was the largest he had ever seen. The Irish Industrial Village employed a total of 105, including 40 from Ireland. In the Village women demonstrated Irish home industries of embroidery, handloom weaving, spinning, knitting, and even butter churning. Visitors could buy Limerick and "Kells" lace the women made as well as bog-oak carving, marble goods, and jewelry. The art gallery in the Village included Begg's painting of "Gladstone Bringing in the Home-Rule Bill," commemorating the bill Gladstone introduced at the beginning of 1893 to give Ireland a parliament. Display of this painting was perhaps calculated as much to please the Irish in America as to twit their English masters across the Irish Sea, who defeated the bill the following year.[5]

Before he left Chicago, Plunkett attended a meeting of the Chicago committee of the association. Although Lady Aberdeen later said Plunkett was "well satisfied" with the efficiency and economy of the project, he wrote in his diary that money had been spent "recklessly." Nevertheless, receipts from the exhibition were strong, and he was sure he would get back his £1,000 investment. Lady Aberdeen later said the exhibition returned a gross profit of about £50,000, of which half went to the exhibition authorities,

---

[4] Plunkett is quoting Voltaire's *Candide*. Plunkett said he subscribed £1,000, which agrees with Lady Aberdeen's listing of contributors to the £6,000 fund raised for the first concession (£500 was secured in Dublin). The second concession was secured on better terms. (Lady Aberdeen makes this Chicago visit a year too late.) Plunkett diaries, 26 Nov. 1892; and Ishbel, *We Twa*, 324.

[5] *Week at the Fair*, 240–41, 244–45; and Ishbel, "Ireland at the World's Fair," 19. The Second Irish Home Rule Bill was introduced in February 1893, and it passed the House of Commons only to be vetoed by the House of Lords.

£20,000 went to benefit workers in Ireland, and the balance was used to set up a "depot" to sell Irish goods in Chicago.[6]

Although Plunkett succeeded in arranging for the all-important financing for the Irish exhibits, Lady Aberdeen was the one in the limelight. After Lord Aberdeen was appointed viceroy to Canada, Lady Aberdeen had the royal standard flown over her cottage in the Irish Industrial Village. Horace had to counsel restraint to the volatile countess on more than one occasion, as when Lady Aberdeen wanted to call the police to recover the royal standard, which some souvenir hunter stole. When she wanted to take the royal standard down each time she left the Village (much as the Queen does when she moves from one palace to another), Horace risked a minor row by objecting. Lady Aberdeen cried, and fortunately the crisis passed.[7]

From Chicago, Horace went to Washington, D.C., to meet with Carey, who was now senator from Wyoming, to prepare him for the annual meeting of the Wyoming Development Company. Plunkett hoped to get the shareholders to agree to place the company on a more business-like footing. Unfortunately, when the shareholders later met in New York, the somewhat rancorous meeting did not result in a plan to raise the money to complete the Wheatland Project. The eastern shareholders still insisted they should not have to subscribe to new bond issues. Plunkett labored to reduce the bickering among the shareholders, urging them to quit trying to gain advantage over each other. He argued the New York group should buy bonds in proportion to their shareholding percentage in order to bring their investments into harmony with those of the original six investors. Although one of the two large New York shareholders agreed with him, another demurred, and of course the controlling shareholders could not impose their will on the minority.[8]

Nevertheless, the financial structure of the company had to be reorganized, as the $250,000 issue of bonds previously issued to

---

[6] Ishbel, *We Twa,* 332; and Plunkett diaries, 7 and 18 Oct. 1893.

[7] Plunkett diaries, 7 and 17 Oct. 1893. The index to the exposition handbook listed the Irish Industries Exhibit as "Lady Aberdeen's," and the descriptive text was signed by her. Lord Aberdeen was sworn in as governor general of Canada (viceroy) on 18 Sept. 1893.

[8] R. Fulton Cutting agreed with Plunkett, whereas Ladenburg, Thalman & Co. were opposed. Plunkett diaries, 20–25 Oct. 1893; Horace Plunkett to Joseph Carey, 23 Oct. 1893; and Wyoming Development Co. minutes, 11 and 19 May 1894.

shareholders was in arrears. In the 1894 reorganization the western shareholders capitulated, and the 1885 bonds were simply cancelled. A new issue of $425,000 of 6 percent bonds was then sold for cash to pay outstanding liabilities, giving the company a sounder financial footing for a time. Although more crises followed, the most serious threat to the company was past, because much of the necessary construction on the project was now finished and the project could begin to attract settlers to the Wheatland area.

By October 25 Plunkett was on the *Teutonic,* headed home. Back in Ireland in November, Beau Watson came to ask him to join in horse dealing venture, with Horace providing his own investment and also advancing an equal share for Watson. Plunkett agreed to this proposal, saying, "I know of no better way of putting that very helpful & satisfactory protégé of mine on his legs." A few days later, Plunkett drew up a new will, naming Watson a co-executor.[9]

Plunkett was elected to Parliament in 1892, and this position took a good deal of his time, both in England and Ireland, making it more difficult for him to fit his customary American visits into his schedule. Moreover, his health, which had always been troubling to him, was declining so badly that he had to go to the Bad Mannheim spa in Germany to recuperate. He took Beau along to care for him. Consequently, for the first time since 1879, Plunkett did not travel to the United States in 1894.

Although Plunkett did not go to Wyoming in 1894, this was the year that the Wheatland Project finally got off the ground. The company followed Plunkett's advice and hired Irad W. Gray of Eaton, Colorado, to sell land in the project. However, a serious management problem arose in the spring of 1894 when Andrew Gilchrist died. Gilchrist had been superintendent of the Wheatland Project since Plunkett first installed him in that position, and now the shareholders had to hire a replacement. Fortunately for the company, Gilchrist's death also coincided with the end of the long development phase for the project. Settlers could now be

---

[9] In February 1892, Beau Watson thought that Countess Maggie Hoyos was in love with him and came to Plunkett to find out whether Horace intended marrying the countess, which Horace did not. Then, in June, Beau was back with the awful news that he had been invited to Countess Hoyos's wedding to Count Herbert Bismarck. Plunkett, Beau Watson, and Watson's widowed sister, Myra Branbury, were to be equal partners in the horse venture. Plunkett diaries, 18 Feb. 1892, 17 June 1892, 22 Nov. 1893, and 2 and 14 Dec. 1893.

attracted, and profitability seemed assured, although profits were still a long way off.[10]

The Wheatland Project opened to settlement in 1894, and the project received some help from by the Carey Act, sponsored by Senator Carey, which became law in the fall of the year. This law was designed to encourage reclamation of arid lands in the West. Through it, up to one million acres would be donated to each state with desert lands, and that land was to be sold to actual settlers. Parcels were to be no larger than a quarter section (160 acres), and the settler had ten years in which to cultivate twenty acres in each quarter section. The law did not specify how the irrigation works were to be built, but in most cases, private parties built them. It was not a particularly successful land law, and in the first five years, Wyoming was the only state with lands developed under it (including some in the Wheatland Project).[11]

By the end of 1895, there were over one hundred farmers on Wheatland Project land, as compared with only fifteen who had lived there the year before. The progress was good enough to make Wheatland the only bright spot that year among Plunkett's American investments. He now could hope that he would eventually recover all of his investment, even though he would never realize an adequate return on the money.[12]

Former Wyoming associates continued to come across the Atlantic to see Plunkett. Boughton came in 1894, with a gloomy view of Wyoming business, and Teschemacher, Alexis Roche, and Alston also came by just to talk. Later in the year, when Plunkett

[10] Wyoming Development Co. minutes, 3 Jan. 1894; and Horace Plunkett to Joseph Carey, 15 March 1894, Carey collection. Plunkett was puzzled by the fact that Gilchrist had started his adult life as a mere private in the army and had built a fortune that at one time amounted to £30,0000 to £100,000 (Plunkett's estimates). In the end, Gilchrist's heavy investments in the cattle industry and land ruined him, and he died insolvent. Plunkett diaries, summary of the year 1894.

[11] Wyoming accepted the Carey Act in 1895. Wyoming was granted an additional one million acres of Carey Act land in 1908, but the total patented under the act in Wyoming was only a little more than 200,000 acres. Although over 31,000 acres were segregated under the Carey Act for the Wheatland Project, only about 3,700 acres were ultimately patented. Carey Act, 28 *U.S. Statutes at Large 422* (18 Aug. 1894); and Robbins, *Our Landed Heritage*, 328–29.

[12] Plunkett diaries, 14 Nov. 1895; and *The Wheatland World*, 1 Nov. 1895. Plunkett researched the possibility of bringing a sugar factory to Wheatland to enhance land values in the project. At the end of 1892, he visited two scientists to discuss the matter, and in 1896, he visited a sugar factory at Norfolk, Neb., that was capable of cutting 350 tons of beets a day.

was in Germany, Teschemacher stopped by again to see him. Boughton was back in the spring of 1895, with the news that he had "caught" a millionairess and planned a July wedding. The next month, Plunkett gave a dinner party for Boughton and his fiancée, as an effort to "qualify" Boughton in the eyes of the prospective mother-in-law. "We did it well!" wrote Plunkett.[13]

When Plunkett came back to New York near the end of 1895, he had his first interview with Theodore Roosevelt, which marked the beginning of a relationship that lasted a number of years. Plunkett's evaluation of Roosevelt was as follows: "A man of clear brain, lofty character, great insight & indomitable purpose." Plunkett was disappointed that his first interview with Roosevelt lasted only fifteen minutes, but it is easy to understand why Roosevelt was not in a mood for long discussions with strangers, for he was then engaged in a fierce political battle.[14]

In the spring of 1895, Roosevelt was appointed police commissioner of New York City, and he was elected president of the commission. He quickly set about rooting out corruption in the police department, starting with the ouster of the police chief. However, when he moved to enforce the state's Sunday Excise Law, which forbade the sale of intoxicating liquors on Sunday, he hit a major obstacle. This law was enacted to please the rural temperance vote, but in the city, with its large Irish and German population, it was largely unenforced; hence, Roosevelt's action was highly unpopular. In the November elections, the Tammany Democrats gained control of the city, and Roosevelt was blamed for this result; thus, his future in the city job was doomed, although his future in national Republican politics was assured. Even though Plunkett's first meeting with Roosevelt was no more than a cordial exchange of greetings, there would be many more meetings of more substance.[15]

Plunkett was a man to be reckoned with in Ireland, where some were even mentioning him as possible chief secretary for that trou-

[13] Boughton's western investments later turned sour, and near the end of 1896, Plunkett and Windsor went to Irvington, Neb., to look at the large farm Boughton had bought there with his brother-in-law, Sir Offley Wakeman. The Windsor & Co. partnership pastured some cattle on the farm, which had now been turned over to Boughton's creditors. Also, Plunkett diaries, 6, 11, and 16 March 1894; 22 May 1895; and 18 June 1895.

[14] Plunkett had a letter of introduction to Roosevelt but did not say who introduced him. Plunkett diaries, 27 Nov. 1895. Roosevelt was ranching in Dakota Territory when Plunkett was in ranching in Wyoming Territory.

[15] Morris, *Rise of Theodore Roosevelt*, 482–514.

bled kingdom, but his name was also achieving recognition in the United States, and not just in the western ranching country. An example of this recognition came early in March 1896, when the editor of the *Boston Globe* asked Plunkett to send a Saint Patrick's Day greeting to its readers; Horace obliged.[16]

In November 1896, Plunkett once again landed in New York, and reporters from six newspapers wanted his opinion on matters in Ireland. His quotable response was that Ireland needed to revive her industries far more than she needed home rule. The next day, he dined with Edwin L. Godkin, editor in chief of the *New York Evening Post.* When Plunkett got to Omaha, the *Omaha Bee* sent a man to interview him. However, Plunkett decided to write the interview for the reporter in order to introduce the idea of agricultural cooperatives in Nebraska and to publicize his own cooperative work in Ireland.[17]

Plunkett's investments in Omaha real estate—mentioned earlier—ultimately amounted to more than $500,000, dwarfing his other American investments. Indeed, the Omaha assets were the largest item left in his estate when he died, and the proceeds from their liquidation proved to be the most significant financial legacy enjoyed by his heirs. However, Omaha land values were not immune to fluctuations in the American economy. Omaha land prices rose sharply in the early 1880s only to plunge beginning in 1887. The ensuing depression in the city was severe, and recovery did not begin until 1898.[18]

The depression in Omaha did not halt new developments in South Omaha, where William Paxton established the Union Stockyards in 1884, acting largely on the urging of Alex Swan. South Omaha Land Syndicate, where Swan was president, platted the city of South Omaha. Although Swan sold out of the stockyards company in 1885, he continued in the land syndicate until his financial empire collapsed early in 1887, at which time he transferred his South Omaha interests to John H. Bosler of Pennsylvania. A key

[16] The *New York Times* also began to take notice of Plunkett, and when he arrived in New York at the end of 1896, a *Times* reporter came to interview him; on subsequent visits, his opinions were often carried in that paper. Plunkett diaries, 7 March 1896.

[17] *New York Times,* 21 Nov. 1896. Edwin Lawrence Godkin was born in Ireland, 2 Oct. 1831, was admitted to the bar and wrote for the *London News,* the *New York Times,* and the *Nation* before coming to the *Post* in 1881. Plunkett declared the *Bee* reporter was an "ass." Plunkett diaries, 20, 27, and 28 Nov. 1896.

[18] Dustin, *Omaha & Douglas County,* 66–67.

individual in these developments was James H. McShane (nephew of John Creighton), who was president of the stockyards in the years 1884–94 and was also a director of the South Omaha Land Syndicate. It is likely that McShane was the man who introduced Plunkett to the South Omaha project.[19]

South Omaha land appreciated rapidly, and parcels that sold for $300 in 1884 brought as much as $7,000 only a year later. Plunkett bought a lot in Omaha from James H. McShane, and Plunkett's partners made other purchases in South Omaha on contracts executed in 1891. In the period of 1900 to 1901, Plunkett bought a few more pieces of real estate in South Omaha, but his outlay for them was probably less than $10,000.[20]

Although the investments were sound, managing his interest in them was a continuing problem for Plunkett. Harry Windsor was his partner and also his agent, and for a time Frank Kemp was also a partner. After a nasty confrontation, Plunkett dismantled the Windsor & Kemp partnership in 1895 and replaced it with a new one that involved only Windsor and himself, suffering an £8,000 loss in the process. Unfortunately, Harry Windsor became bankrupt the following year, imperiling the new partnership, where the $50,000 debt to Plunkett was the largest on the books. In the settlement dissolving the partnership, Windsor transferred half a dozen parcels of real estate to Plunkett. Even though their partnership was now dissolved, Plunkett kept Windsor on the payroll as his agent in Omaha, in charge of his business in that region.[21]

On his return to New York at the end of 1897, Plunkett met a man who had great interest in Irish matters, a man who was to become a lasting friend. William Bourke Cockran, who soon became Plunkett's "new Irish friend," had a luxurious flat in New York City and

[19] *Omaha Stockyards*, 2. John H. Bosler was the uncle of Frank Bosler, who was later Plunkett's partner in Wyoming.

[20] Two parcels in 1900 cost less than $2,500, and four others in 1901 cost $5,600; there was a third purchase where the consideration is uncertain. Register of deeds, Douglas County, Neb. South Omaha was annexed to the city of Omaha in 1915 after a rancorous debate in which the packing houses took the side of annexation and pressured their employees to support it. *Omaha Stockyards*, 15.

[21] In Chicago, Plunkett met with John Clay, Jr., who gave him additional information about Kemp's dealings through the commission firm of Clay & Forrest, convincing Plunkett that Kemp's "robbery" of him was more than he had thought. Plunkett diaries, 20 and 21–23 Nov. 1895, 25 Nov. 1896, and 15 Dec. 1896. When Plunkett further reorganized his Neb. holdings, he placed the Hiland farm in a corporation and had Johnny Peirce, his old WP foreman, come to manage it at the end of 1896.

a country home on Long Island, near Port Washington. It is uncertain how the two men were introduced; it may have been through Cockran's Irish connections (he was born there) or as the result of Plunkett's friendship with Harry Oelrichs, the one-time Wyoming rancher, whose brother knew Cockran.[22]

Cockran's roots were very different from Plunkett's, which may have heightened Horace's early interest in Cockran, for Horace was always interested in social strata other than his own. Cockran started as a barefoot lad in Sligo, Ireland, but by the time Plunkett met him, he was a rich man who was deeply embroiled in New York politics. Cockran was very charismatic, and Plunkett was quoted by Cockran's biographer as having said that Cockran "radiated vitality and charm," and that when he entered a room, it was like turning on a light.[23]

Characteristically, Plunkett analyzed Cockran, saying, "He is evidently rich, but he has a scheme of life which is unAmerican & most laudable." Later, he prophesized that Cockran's great oratorical skills would lead him to a fine career in politics. He sensed that Cockran was really religious but struggling between his duty to that side of his character while at the same time enjoying the fruits of his wealth. Plunkett wrote, "I left him thinking, 'What shall I do to be saved?'" and then supplied the answer he wanted: 'I hope he will serve Ireland in America.'" Plunkett's agenda for Cockran was to educate him on the Plunkett plans for Ireland in the hopes that Cockran could bring helpful Irish American influence to bear on Plunkett's work there.[24]

From New York, Plunkett went to Washington, D.C., where he received blue ribbon treatment. On December 6, he witnessed the opening of Congress, and two days later he met Boss Thomas C. Platt of New York, who was now a U.S. senator. Platt's health was

[22] When Bourke Cockran was a struggling lawyer (and gifted orator) in New York, he joined the Hyena Club, which admitted only those whose net worth did not exceed $30. Exceptionally, Harry Oelrichs's brother Herman was admitted to the club on the understanding that he would pay all of the club expenses. By the time Plunkett met Cockran in 1897, Cockran had retired from Congress (he was in the House from 1891 to 1895) and become a wealthy lawyer—earning fees of $100,000 per year—whose home was always open to his friends. McGurrin, *Bourke Cockran*, 34, 94–95, 115. Also, Plunkett diaries, 3 Dec. 1897. Harry Oelrichs, who was born in 1854, was the manager of the Anglo-American Cattle Co. His brother Charles May Oelrichs also ranched with Harry in the West, and their brother Herman ran Oelrichs & Co, shipping merchants in New York.

[23] McGurrin, *Bourke Cockran*, 334.

[24] Plunkett diaries, 22 Nov. 1898, 5 Dec. 1901, and 16 Nov. 1902.

declining, and he contributed little to the Senate's work, but off the floor he controlled political patronage for New York, making weekly trips to New York City to keep in touch with the public pulse. Platt took Plunkett to the Senate and arranged for him to have the privilege of the floor, where he met Sen. William V. Allen of Nebraska, the Populist maverick whose 1895 filibuster of nearly fifteen hours was a record for the time. A visit to the Library of Congress rounded out the day, and the next day Plunkett went to the Agriculture Department and later visited with the Speaker of the House, Congressman Thomas B. Reed of Maine. This visit to official Washington clearly testified to Horace Plunkett's increased stature in the United States.[25]

From Washington, Plunkett went to Omaha to see Windsor and Chaplin, and then he went to Cheyenne to see Carey on Wyoming Development business. Carey went with him to Wheatland, where they spoke to a group of 100 farmers. Plunkett talked about the possibilities of cooperatives, and Carey lectured them on their relationship with the development company. Neither of these messages dealt with the company's perennial cash shortage that was severe enough to threaten shutting down the project.[26]

Although the advertising for the Wheatland Project promised settlers there were "no chances of failure," it was apparent there was not enough water to go around. To remedy this situation, it was necessary to construct a new 6,600-acre reservoir at a cost of about $119,000. Work could not begin until the money was raised, even though the farmers agreed to supply labor for construction. To meet this new financial crisis, the Wheatland Industrial Company (hereafter the Industrial Company), with a capitalization of $200,000, was formed to secure financing for the reservoir. Plunkett was a shareholder of the new company and John Chaplin a director.[27]

The Western Union Beef Company, successor of the American

---

[25] Thomas Collier Platt was born in Owego, N.Y., on 15 July 1833, and he served in the U.S. House of Representatives and was twice elected to the Senate. Gosnell, *Boss Platt and His New York Machine*, 172–80. Sen. William V. Allen, elected to the Senate on the Populist ticket, engaged in a number of obstructionist tactics, but his 15-hr filibuster record fell to Robert La Follette in 1908. *New York Times*, 20 and 21 Dec. 1895, 23 April 1897, and 11 July 1937; and Plunkett diaries, 5–9 Dec. 1897.

[26] Plunkett arranged for Windsor to sell Wyoming Development lands on commission. Plunkett diaries, 12–24 Dec. 1897.

[27] The capitalization of Wheatland Industrial Co. was increased to $1 million in 1907. *Platte County Record-Times*, 2 April 1980; and *Wheatland World*, 2 Dec. 1898.

Cattle Trust, was still alive after having struggled for years without notable success. Management changed once again as George Baxter became president of the company, and William Ricketts replaced him as Wyoming manager. Ricketts made his headquarters at the company's Half Circle L ranch in Crook County. Finally, at the end of 1897, Tom Sturgis told Plunkett that the Western Union Beef Company was to be liquidated, and Plunkett hoped to receive some £5,500 from it.[28]

Instead of a quick liquidation, the Western Union Beef Company went through another transformation. In the spring of 1898, the company sold part of its properties, made a $680,000 payment to shareholders, and then changed its name to the Western Live Stock and Land Company. The following year all the cattle in Wyoming and Montana were sold, and a second payment of $306,000 was made to shareholders. These payments amounted to about $7.25 per share, and final liquidation yielded less than an additional dollar per share—on shares nominally worth $100. Thus, the saga of the American Cattle Trust finally ended.[29]

Despite his difficulties with the American Cattle Trust and its successor organizations, Plunkett did not carry a grudge against Sturgis. In the fall of 1913, Sturgis and his wife came to see Plunkett in Ireland. "He has his faults & she rather bores me," Horace wrote in his diary, "but I have known them for 34 years, & they have gone through hard times. . . . I feel kindly toward them."[30]

Plunkett's last entry of the year 1897 recorded the receipt of $100 for his article, "The Irish Question in a New Light," which was published in the January 1898 issue of the *North American Review.* He appealed to the American audience for support in his efforts to emphasize economic growth initiatives for Ireland, rather than

---

[28] Osborne, "John Alexander Osborne," 313; and Plunkett diaries, 28 Dec. 1897.

[29] At the time of the first liquidation payment, Ricketts bought the Half Circle L Ranch from the company and renamed it Sunnyside Ranch. When Ricketts disposed of the Sunnyside Ranch, it was renamed the Padlock Ranch. George Baxter was president of the company at the time of the name change, and Charles T. Leonhardt was secretary; these two men apparently continued in these positions until final liquidation. Plunkett diaries, 3 Dec. 1897. Based on the 136,000 shares outstanding before liquidation began, the two payments of $986,000 amounted to $7.25 per share. In the summer of 1910, the Western Live Stock and Land Co. still owned a $97,000 note due in 1911, plus some lesser notes and receivables. Two smaller liquidation payments were expected, the first one in May 1911. James Byrne to Horace Plunkett, 7 July 1910, Plunkett collection, Library of Congress.

[30] Plunkett described Mrs. Sturgis as "a brainless garrulous *good* woman." Plunkett diaries, 18 and 20 Aug. 1913.

political solutions. Plunkett did call for some subsidies to Irish agriculture and industries, but he primarily advocated cooperative efforts. He dismissed the political agenda with these words: "I believe our chief offence is that we despise that so-called love for Ireland which is but a thinly disguised hatred for England. Our hopes for the regeneration of our country do not involve the destruction of an empire which Irishmen have taken a leading part in building up, and are to-day foremost in maintaining. Such a perverted patriotism is alien to the character of the Irish people, who are neither revengeful nor wanting in intelligence."[31]

In November 1898, Plunkett sailed again for New York and was in Wyoming by the end of the month. He and Carey went to Wheatland, where the farmers were suffering from the water shortage in the crop year just ended, and it fell to Plunkett to deal with their anger. He opened the meeting and explained what the company was doing to increase water storage capacity, while Carey confined his remarks to assurances that the company would persevere. Plunkett felt he had quieted the mood of the farmers, and he groused in his diary, "It is rather annoying that I have to come out from Ireland to settle Carey's muddles."[32]

The year 1899 was another in which Plunkett did not go to America, this time because he had broken his leg. The doctor examined the fracture using X-rays, which had been discovered in 1895 by Wilhelm Conrad Roentgen and had quickly been adapted to medical use. Unfortunately, the early practitioners knew nothing of the risks of overexposure, and in an extreme case Plunkett's first treatment was for fifteen minutes, with a later X-ray reduced to four minutes when a more powerful machine became available. The X-ray treatments ultimately were more dangerous to Plunkett's health than the broken bone had been.[33]

In the fall of 1900, Plunkett was well enough to travel again, but

---

[31] Plunkett, "Irish Question in a New Light," quote at 120.

[32] Plunkett diaries, 29 Nov. 1898, 5 Dec. 1898, and summary at the end of 1898. Also, *Wheatland World,* 2 Dec. 1898.

[33] Plunkett's first accident was on March 13, and he fell and refractured the femur on June 13. The doctor examined the bones with X-rays on June 18 (with a 15-min. exposure) and again on June 24, July 31, September 22, and December 11, when a stronger coil was used, cutting the exposure to 4 min. Plunkett diaries, 1899. In an extreme case of overexposure to X-rays, in 1896 a man's head was exposed to X-rays for 14 hr. He suffered severe burns. Brecher and Brecher, *Rays: A History of Radiology,* 83–84. Maximum exposures for X-rays were finally proposed after the National Council on Radiation Protection and Measurements was established by Congress in 1925.

for the first time in eighteen years he became seasick on the way to New York. When he got to Cheyenne, he found much had changed, for the Cheyenne Club no longer served food, and for the first time in twenty years he checked into a hotel. One noteworthy thing that happened, though, was that George Bosler, who was John Coble's partner, introduced himself to Plunkett, which marked the beginning of an association with the Boslers that lasted a number of years."[34]

Next, Plunkett went to the Wheatland Project, where he and Carey looked at the new dam across the Laramie River, and Plunkett also learned he would have to advance another £3,000 for the project. The big reservoir project was going forward, but in granting the permit the state engineer required that the company develop even more land. This requirement proved very expensive to the company. To satisfy the requirement, the company filed on two tracts under the Carey Act, one for 10,000 acres and the second for 20,000 acres.

A sixteen-and-a-half-mile canal had to be built to develop the 10,000-acre tract. The canal was built through rough, rocky country, at a cost of $67,000, making the initial cost per acre on this tract 60 percent higher than for other lands in the project. The soil was also poorer, and ultimately only 3,000 of the 10,000 acres were developed, at a cost more than four times the average for the rest of the project. The 20,000-acre tract was a complete disaster, and the company spent $162,000 on canals before learning that there was not enough water available to serve these lands.[35]

A second reservoir had to be constructed at a cost of nearly $118,000, but this investment finally guaranteed the success of the Wheatland project. The new reservoir began to fill in 1903, and that year the project enjoyed an adequate water supply, even though the reservoir was not yet full.

Plunkett heard of the possibility that sugar beets could be grown on the Wheatland Project, and in 1902 he stopped at Ames, Nebraska, to inspect a processing plant that cost a million dollars. The managers admitted that profitability was "problematical," and nothing came of the idea at the time. Nevertheless, the idea of rais-

---

[34] Plunkett diaries, 22 Nov. 1900 and 6–9 and 17 Dec. 1900.

[35] The Wheatland Project was developed at $5 per acre, and the canal inflated the cost for the first parcel under the reservoir (the Bordeaux tract) to $8 per acre; when only 3,000 of the 10,000 acres could be developed, the cost became $22 per acre.

ing sugar beets in the Wheatland Project was a sound one, although it took years of planning before the factory could be built.[36]

In 1901, Plunkett made his usual hurried trip to Omaha and Cheyenne. Before Christmas he went to Washington, D.C., where Theodore Roosevelt had succeeded to the presidency in September upon the death of President McKinley. Plunkett met Agriculture Secretary James Wilson, who took him to see the president, and the president asked him to lunch. In a "delightful family party" with Mrs. Roosevelt and the four children, the men talked about the Boer War and Ireland.[37]

In the summer of 1902, Dr. Hollis Burke Frissell of the Hampton Institute in Virginia, who had learned of Plunkett's work in Ireland, came to see Plunkett in London. Frissell wanted to see if Plunkett's ideas for helping Irish farmers could be applied to black farmers in the American South. Plunkett, who harbored at least some racial bias against blacks, was not pleased when Frissell compared them with the Irish, saying, "The Negro is not Irish, but both are more or less human." Despite this initial disdain, Plunkett became interested in the education of black students, and when he went to America at the end of 1902, he met Booker T. Washington, Hampton Institute's most famous graduate, and was impressed. Afterward, he traveled to Hampton to see the institute for himself.[38]

This interesting school, which was founded in 1868 after the Civil War, was a coeducational institute established primarily for black children. It was reasonably well financed from a state grant and an endowment of over $1 million, yielding an annual income of $170,000. The faculty consisted of eighty-two instructors, and when Plunkett visited the institute, the student body included about 700 blacks and some Indians, aged sixteen to nineteen. Male students worked on the school's 600-acre farm to pay their board bill of $10 a month, and female students sewed and cooked. The student's life was not an easy one, for the work was hard and the hours long, and during the first year there was only a little time for school at night. Plunkett left the school convinced that the cooperative movement could apply to the black farmer in the United States.

---

[36] Plunkett diaries, 23 Nov. 1902.

[37] Roosevelt assured Plunkett that he intended to leave other countries' affairs alone, a promise he later had great difficulty keeping. Plunkett diaries, 23 Dec. 1901. President McKinley died on 14 Sept. 1901.

[38] Frissell was born in New York, 14 July 1851. Plunkett diaries, 17 and 21 Dec. 1902.

Plunkett continued to be in demand as a speaker in the United States. He addressed the Omaha Real Estate Exchange in December 1902, discussing improvement of Omaha's development. In New York, Cockran arranged numerous dinners and other affairs to introduce prominent Irish Americans to Plunkett, who was always trying to develop financial support for his Irish work. When Plunkett needed stenographic and other office facilities, Cockran offered his own. These courtesies were a great help to Plunkett, but Cockran's public style was very different from Plunkett's, and he could be a source of irritation in the public arena.

In particular, Cockran's view of the British presence in Ireland was much harsher than Plunkett's. In December, both Plunkett and Cockran were invited to address a political science group at Columbia University, where about 250 assembled to hear them give an overview of the Irish situation. To Plunkett's disgust, Cockran chose to speak on the history of English mismanagement of the island, "which he grossly exaggerated & distorted."[39]

In the East, Plunkett spoke four times, including the speech at Columbia, hoping to raise money for his work in Ireland. At Harvard, he appeared before a "very poor audience" and then went back to New York to speak before the League for Political Education ("no possibly subscribing people present"). His reaction to all the attention was, "The Americans kill me with hospitality. They were nice, but crudely cultured people."[40]

The British government never officially recognized Plunkett's extensive public work, perhaps in large part because of his unfortunate penchant for irritating his friends as well as his enemies. In the spring of 1903, Lady Aberdeen confided to Plunkett that she had asked Prime Minister Arthur J. Balfour to have Plunkett ennobled, but nothing came of that effort. However, later in the year, the King knighted him, under circumstances fraught with almost comic difficulties.[41]

The occasion arose when the King and Queen came to Ireland

---

[39] Plunkett diaries, 12 Dec. 1902.

[40] A critical comment on the Plunkett lectures appeared in the *New York Times* on 21 Dec. *New York Times*, 13, 17, and 21 Dec. 1902; and Plunkett diaries, 3, 5, 12, 15, and 16 Dec. 1902.

[41] Plunkett diaries, 24 March 1903. After noting Lady Aberdeen's conversation, Plunkett wrote, "I wish she had not." Arthur James Balfour was named prime minister on 11 July 1902, succeeding his uncle, Lord Salisbury.

on a state visit, and Plunkett, who was still in the government, assisted with preparations for the royal visit. At the end of visit, the royal party left Queenstown for Cork, and Horace, believing his official duties to be at an end, left on the train for Dublin, only to be awakened by a guard who had a wire stating that the King wished to see him. He went to Cork to board the royal yacht, but the only clothes he had with him were a very dirty woven coat, a yachting cap, and a "reefer" suit—certainly not suitable for presentation to the King. At 2:30 A.M., all the shops in Cork were closed, so when Plunkett boarded the royal yacht, Lord Dudley had to lend him a frock coat. Attired thus, Plunkett was then taken into the royal presence, still wearing dirty boots.

The King made a short speech in which he thanked Plunkett for his service to Ireland and for kindnesses extended to the royal couple, and then he said that he was giving Plunkett the second class of the Victorian Order. Horace was not the only one unprepared for the ceremony, as a cushion had to be found for him to kneel on, and the King borrowed a sword for the dubbing. It was "most informally performed," as Plunkett later wrote. Afterward, Sir Horace rose and chatted with the King about Ireland. Lord Dudley, who was then viceroy, also emphasized to him that the King wanted him to understand that the honor was a personal gift and was not given on behalf of the British government.[42]

Plunkett's friends in the American West tried to recognize the change in Plunkett's status, albeit not always correctly. In Omaha, someone called him "Sir Plunkett," and although the reporter from the *Omaha Bee* knew that Plunkett was properly "Sir Horace," he nevertheless mistakenly called Plunkett a peer, rather than a mere knight. Plunkett responded that he much preferred to be called "Mr. Plunkett" but acknowledged the American fascination with titles, adding, "I believe it is Mark Twain who said that of the 1,500 colonels and generals alive at the close of the Civil War, there are now only 15,000 surviving."[43]

---

[42] Plunkett diaries, 21 and 22 July 1903 and 1 Aug. 1903. The Royal Victorian Order was established by Queen Victoria on 21 April 1896 as a personal gift of the sovereign to those who performed personal service to the sovereign; the awards were paid for by the sovereign. Plunkett was dubbed Knight Commander of the Victorian Order, the second of five classes. Edward VII was fond of the order, giving it frequently, "almost as some kind of a tip." Vickers, *Royal Orders*, 117.

[43] *Cheyenne Daily Leader*, 26 Oct. 1907, quoting the *Omaha Bee*.

10

Advisor to a President

AT THE BEGINNING of December 1903, Plunkett once more landed in New York. For some reason, this time he did not stay at Bourke Cockran's apartment; instead, Cockran booked rooms for him in the Waldorf–Astoria. When it opened in 1897, the one-thousand-room hotel was the largest in the world, with a ninety-five-foot ballroom and three hundred-foot marble corridor, making it the most elegant and the most expensive in the city. Although Plunkett growled that it was a "monster," still he was also impressed that it was a marvel of organization.[1]

Plunkett went to Washington, D.C., for lunch with the president before going to Omaha to look after his personal business. That morning U.S. Marines had landed in Panama as a show of force in the area, and President Theodore Roosevelt chatted with Horace about the situation there. Earlier in the year, Panama had declared its independence from Colombia and then promptly signed a treaty permitting the United States to construct a canal across the isthmus. The president shared some confidences about the situation with Plunkett, who did not set them down in his diary.[2]

---

[1] The Waldorf–Astoria was actually two hotels: the Waldorf, owned by William Waldorf Astor, and the Astoria, owned by his cousin, John Jacob Astor IV. The two hotels were joined by corridors that could be sealed in the event of a familial falling-out. Kaplan, *When the Astors Owned New York*, 79, 84–88. Also, Plunkett diaries, 10 Dec. 1903.

[2] The United States was concerned that Colombia might use force to reclaim Panama, and while the situation was unfolding, Roosevelt kept U. S. naval forces in the area. Roosevelt sent a "cruise" of four battleships to the West Indies at the beginning of Dec., a company of marines landed in Panama a week later, and a second company landed on 14 Dec. *New York Times*, 2, 9, and 15 Dec. 1903; and Plunkett diaries, 14 Dec. 1903.

In Omaha and Cheyenne, Plunkett spent two days on his own business before returning to Washington and then going to Richmond, where he wanted to learn whether tobacco could be grown in Ireland. On the way north to Philadelphia on the train, he met Frank C. Bosler, who wanted to interest him in investing in a ranch near Rock River in Wyoming. Plunkett may have met Frank Bosler earlier, and in any case this was no chance encounter, for Windsor and his partner from Omaha had already optioned a one-third interest in the Bosler ranch. Windsor was still Plunkett's agent in Omaha, but he was still in financial difficulties, so it is obvious that Plunkett's resources were needed to give substance to the option.[3]

The opportunity to get back into the western ranching business immediately attracted Plunkett, particularly since the property was fenced fee land, rather than government open range, and should have fewer risks than open range ranching. A second motive for Plunkett to invest was the fact that Windsor, who would manage the ranch, owed Plunkett money that he could not repay. Thus, Plunkett hoped that if the Diamond Cattle Company were profitable, Windsor could make at least partial payment on his debts. A final incentive for Plunkett was the thought that this might be an opportunity to introduce his twenty-five-year-old nephew Eddie, the new Lord Dunsany, to the ranching business. Harking back to his own life on the range, he hoped that the rigors of ranch life might promote a healthier lifestyle for Eddie than that of poet and playwright.[4]

Unfortunately, the Diamond Ranch was a project Plunkett came to hate. Although a number of his other ventures failed to earn a

---

[3] Plunkett diaries, 22–28 Dec. 1903. The four original Bosler brothers were James Williamson, who was the spokesman for the partnership until his death in 1883, George M., Abram, and John Herman. Frank C. Bosler, who was born in May 1869, was James W. Bosler's son. The Bosler Brothers had invested in Iowa in the 1850s, but their cattle interests elsewhere were very extensive, and they were heavy beef contractors for the government in the West. J. Gardiner Haines, a grain dealer, was Windsor's Omaha partner.

[4] Plunkett's brother John William Plunkett died on 16 Jan. 1899 and was succeeded as Lord Dunsany by his son, Edward John Moreton Drax Plunkett, who was born 24 July 1878. When Plunkett introduced Dunsany to the Bosler ranch investment, Dunsany was writing plays and poetry but had not yet been published. Indeed, he had to pay to have his first book published. However, over the next fifty years, he successfully published more than ninety works, some of which have been reprinted several times.

profit, for the most part he enjoyed the experiences they gave him. Although he plunged into the Diamond Ranch project with enthusiasm, Plunkett's relationship with Frank Bosler eventually soured to the point that he wanted nothing more to do with the man. Plunkett found himself trapped in the investment and was not able to exert his customary influence over management to improve the chance of recovering his money.

By the time Plunkett returned to the United States at the end of 1904, the Diamond Cattle Company was organized with a $215,000 capitalization, consisting of property contributed by the new shareholders. Windsor and his Omaha partner sold their equity in the Cooper Ranch on Rock River to the company for $149,100, and the Diamond Cattle Company assumed the $100,000 mortgage against the property. Bosler transferred a total of 2,650 head of cattle to the company, and Plunkett contributed his cattle and horses on the Rock Creek ranch, allotting his investment to his nephew, Lord Dunsany. The shareholders of the company increased the capitalization to $300,000, requiring the first of many additional investments by Plunkett. In the four days Plunkett spent on the ranch, the ranch fever returned to him, and he spent a day and a half surveying a reservoir site to store irrigation water for 5,000–7,000 acres.[5]

Although he would still encounter many difficulties with his American investments, Plunkett finally had reliable men to look after his interests there. Conrad Young was one of these men, and Plunkett hired him to work in the Omaha office in 1904. To his surprise, when he wanted to give Young more responsibility, Windsor objected, saying that Young was dissipated, weak, and untrustworthy, having "practically" robbed the safe some years before. Fortunately, Plunkett did not let this strange negative

[5] J. Gardiner Haines was Windsor's partner in the Cooper Ranch. In 1905, Bosler acquired Haines' interest in the Diamond Co. The Diamond Co. shareholders were Bosler, with 392 shares; Haines, with 775; Windsor, with 716; Plunkett, with 133; and Lord Dunsany, with 134, for a total of 2,150; hence, Plunkett's share was 12.4 percent of the total. Minutes of the Diamond Cattle Co., 4 Nov. 1904, Bosler Collection, American Heritage Center, University of Wyoming. The Diamond Cattle Co. did not include other Bosler properties in the area. The old Ione ranch that Boughton had managed was resold to Frank C. Bosler in 1900, when it became a part of the 50,000-acre Iron Mountain Ranch, which was owned by John Coble and Bosler at the time but later was owned almost entirely by the latter. Also, Plunkett diaries, 28 Dec. 1903 and 4 Feb. 1904.

assessment deter him, and Young became a source of strength in Omaha that extended beyond Plunkett's death.[6]

Plunkett tried hard to obtain a measure of control over the Diamond Company. He insisted that his own auditor examine the books of the company. He installed his trusted friend and ex-foreman, Johnny Peirce, as foreman of the Diamond Company cattle operations. Finally, he expected that John Chaplin in Cheyenne would also be able to keep an eye on the project and help him to control the direction the company would take. Although Chaplin was always good at keeping Plunkett informed, any hope that he could control Frank Bosler was wildly optimistic.[7]

Windsor's association with the Diamond Ranch was brief, for he died in the spring of 1906. Windsor's death was doubly troubling to Plunkett. Initially, he was concerned that Windsor's death left the entire management with Bosler, whose capability was untested. But later Plunkett also learned that Windsor had not been the faithful steward he had supposed. Because of his concerns about Bosler, Plunkett now felt obligated to take over his nephew's investment, and after acquiring these shares, Plunkett owned 15.35 percent of the Diamond Company. This was a substantial commitment to the company but not large enough for Plunkett to be able to force Bosler to accept his point of view. Extracting Lord Dunsany from the Diamond Ranch not only spared the young baron the task of dealing with Bosler, but also deprived Plunkett of a reason to divert Dunsany from becoming a successful and prolific author.[8]

[6] Conrad H. Young was born in England in January 1875 and came to the United States in 1892. His obituary states that he went to work for Plunkett in 1888, which cannot be correct. In a letter to Young in the fall of 1926, Plunkett mentioned that Young had been working for him for 36 years. *Omaha World-Herald,* 28 March 1962. Young died 27 March 1962. Also, Plunkett diaries, 13 Dec. 1904.

[7] Johnny Peirce remained in the foreman's post until he died in 1917; he also had a small stock ownership in the company. Plunkett diaries, 4 Feb. 1904. Thomas C. Cannon of Omaha was Plunkett's auditor. Born in England in 1866, Cannon came to the United States in 1889 and practiced accounting in Omaha until 1908, when he moved to Spokane, Wash., and opened an accounting practice there. He continued to serve Plunkett from Spokane.

[8] Plunkett diaries, 28 April 1906. Windsor's widow continued to hold shares in the Diamond Co., and there were also other shareholders not related to the Plunkett and Bosler interests. Frank C. Bosler to Horace Plunkett, 9 March 1912, 6 Oct. 1912, and 1 Jan. 1914, Bosler collection, American Heritage Center, University of Wyoming (hereafter cited as Bosler collection).

After visiting the Diamond Ranch, Plunkett stopped in Laramie, where he visited the University of Wyoming, a struggling school that had been established in 1886. His host was the president of the university, Fred M. Tisdel (only in his position since July), who assembled the student body to hear Plunkett speak. Tisdel was the most highly qualified president in the early history of the university, holding three graduate degrees, including a doctorate from Harvard, but his greatest qualification in Wyoming was the fact his uncle was Sen. Clarence D. Clark. The university was dominated by a Republican board and its secretary, Grace Raymond Hebard, who wielded nearly unquestioned power over university finances and appointments.

Plunkett gained the impression that the university was mostly devoted to agriculture and mechanical arts, undoubtedly because federal funding (providing 80 percent of university financial support) was supposed to be devoted to those subjects. In practice, Miss Hebard financed the English and Education departments from this source as well as a new floor in the physics department, to name only some of her funding irregularities. Nevertheless, Plunkett was correct in his criticism that the liberal arts courses were "thinly" attended (the liberal arts curriculum had fewer than 20 students), and also there was no law or medicine. What he did not say was that almost half of the two-hundred-person student body was enrolled in high-school-level courses. Plunkett noted that over half of the students were women, and he was impressed with the exciting spirit of this "citizens" university, calling it "an interesting beginning." By comparison with his own educational background, he concluded that the Wyoming school was not really a university but was the best that could be provided by people "who couldn't afford a [real] university."[9]

Continuing his visits to educational institutions, Plunkett stopped in Ames, Iowa, to visit the Iowa State College of Agriculture and Mechanic Arts, which was a land grant college like the University of Wyoming. Next, he went to Chicago University and finally back to Washington, D.C., for lunch with the president and an interview with Maurice Low, correspondent for the *London Morning Post.* In New York again, Plunkett had lunch with Peter

[9] Hardy, *Wyoming University*, passim; and Plunkett diaries, 8 Dec. 1904.

Collier, publisher of *Collier's Weekly,* which was then one of the largest magazines in the United States.[10]

Another example of the wide range of Plunkett's interest was an experiment being conducted at Wesleyan University by Professor Wilbur O. Atwater. Plunkett hoped the experiment would be useful to his work in Ireland. Atwater had constructed a respiratory calorimeter chamber large enough to contain a man for experiments that ran from a few hours to two weeks. This chamber was used to measure accurately how much heat a subject gave off while sitting, lying down, and doing exercises, thus indicating how many calories were being burned. Because half the working family's income was spent on food, Atwater wanted to teach the poor what sources of protein were most economical. Although much effort was expended to be certain that the tests were valid, Plunkett noted, "So far, results are few."[11]

The year 1903 was a good one for the Wyoming Development Company, and Carey was moved to declare to the shareholders that the colony's future "may no longer be considered a question." In some cases, farmers realized as much as $50 per acre from their farms, even with "indifferent" improvements, and the improved economic conditions justified raising the price of land still for sale by 25 percent or even 50 percent. The following year, Horace permitted himself an optimistic conclusion in his diary: "It is a terribly long-winded investment, but I think if we hold our health long enough, our children would have a fine inheritance." After the annual meeting at the end of the year, he added, "The company is at last in smooth water & my stake in it is large & will be ultimately remunerative—this after 22 years."[12]

At the end of his 1904 journey to America, Plunkett spent a good

---

[10] Plunkett's diaries, 15–22 and 26 Dec. 1904. Collier was a social reformer who was often sued, and his journalistic style was adopted by others who formed the group Theodore Roosevelt later referred to as "muckraking" journalists. The term "muckraker" was based on the character Man with the Muckrake in John Bunyan's *Pilgrim's Progress.*

[11] Plunkett diaries, 23 Dec. 1904. Wilbur Olin Atwater received the PhD degree from Yale, and his research was heavily influenced by German nutritionists. In 1903, he was named chief of nutrition investigations in the U.S. Department of Agriculture (USDA). Atwater's room calorimeter was sensitive enough to measure the rise in room temperature when the volunteer inside adjusted his watch in the middle of the night. The calorimeter, in a less claustrophobic design than Atwater's, is still used at the USDA Research Center in Beltsville, Md.

[12] Joseph Carey to Stockholders of the Wyoming Development Company, 1 Jan. 1904, Carey collection. Also, Plunkett diaries, 9 and 27 Dec. 1904.

deal of time with intellectuals talking about the cooperative movement and also discussing and defending his newly published book, *Ireland in the New Century*. In a section widely criticized by the Roman Catholic Church, Plunkett gave an evaluation of the influence of religion on Irish society, in which he spared neither the Protestants nor the Catholics:

> I find that while the Protestants have given, and continue to give a fine example of thrift and industry to the rest of the nation, the attitude of a section of them towards the majority of their fellow-countrymen has been a bigoted and unintelligent one. On the other hand, I have learned from practical experience amongst the Roman Catholic people of Ireland that, while more free from bigotry, in the sense in which the word is usually applied, they are apathetic, thriftless and almost non-industrial, and that they especially require the exercise of strengthening influences on their moral fibre.[13]

These opinions provoked negative responses from both sides of the issue, both in Ireland and among the Irish community in the United States.

When he visited President Roosevelt at the beginning of 1906, Plunkett began to hope that his work on rural economic problems in Ireland might well be transplanted to the United States, as well. The president was at once interested in the idea and asked Plunkett to write down his views on the subject, which he did after returning to Ireland. Again, at this meeting Roosevelt passed on some confidences to Plunkett, which Plunkett did not write about in his diary, saying, "I cannot enter here the most interesting thing he said to me."[14]

Plunkett sent his paper to Roosevelt in June, and in the middle of July, the president replied to Plunkett's letter. Writing to "My Dear Horace," the president said, "By George, I wish you were an American and either in the Senate, or my Cabinet. You take an interest in exactly the problems which I regard as vital, and you approach them in what seems to me to be the only sane and healthy way."[15]

At the end of 1906, Plunkett was back in the United States, and

[13] Plunkett, *Ireland in the New Century*, 120–21.

[14] Although Roosevelt asked Plunkett to write down his rural life views in January, Plunkett did not send the paper until near the end of June. Plunkett diaries, 12 Jan. 1906 and 24 June 1906.

[15] Theodore Roosevelt to Horace C. Plunkett, 3 July 1906, Theodore Roosevelt collection, Library of Congress.

when he arrived there was a message from Roosevelt's private secretary inviting him to dinner at the White House. The other guests had interests and responsibilities in areas that also interested Plunkett. They were Gifford Pinchot, chief of the Forest Service (whom Plunkett called a "Frenchy"); James Rudolph Garfield, son of the late president, who had been nominated as secretary of the interior; and Oscar Straus, who had just been confirmed as secretary of commerce and labor. Roosevelt began the conversation with an effusive apology to Plunkett for plagiarizing his words for the annual message to Congress.[16]

Plunkett said there was "much talk" at the dinner, and the party became "ultra indiscreet." The first topic Plunkett discussed was the Bellamy Storer incident, which the newspapers had reopened a week before, raising questions about Roosevelt's reason for recalling Storer as ambassador to Austria. The matter was extremely sensitive for Roosevelt because he had a close personal relationship with the Storer family. The Storers had been neighbors of the Roosevelts in Washington, D.C., when Storer was a congressman and Roosevelt was a member of the Civil Service Commission. Storer was also godfather to one of the Roosevelt boys, and Mrs. Storer's nephew married the president's daughter Alice.

Mrs. Storer was wealthy and very involved in politics. Archbishop Ireland of St. Paul converted her to Roman Catholicism, and she and the archbishop supported McKinley's election to the White House. Mrs. Storer said that after McKinley was elected, she asked him to appoint Theodore Roosevelt as assistant secretary of the navy, and McKinley did so. McKinley also appointed Bellamy Storer as minister to Belgium (he was transferred to Spain later). After the Spanish–American War, Archbishop Ireland was appointed to the Philippine Commission and became involved in the negotiations that resulted in purchase and redistribution of

[16] Gifford Pinchot's father was born in France, but Gifford was born in Conn. In 1896 he was named chief of the Division of Forestry in the Department of the Interior, and when the Forest Service was created in the USDA, he was the first chief. In his long message dated 3 Dec. 1906, Roosevelt called attention to the need to improve the farmer's condition through education "in the widest possible sense" and pointed out the desirability of "the association of farmers for mutual advantage." He also noted the desirability of cooperation between the federal government and the states. All of these ideas were, of course, on target with Plunkett's pronouncements, and Plunkett noted that the president had not only adapted his ideas, but also "some of the sentences." Plunkett diaries, 14 and 16 Dec. 1906 and summary for the year 1906.

church lands in the Philippines. Although Roosevelt was grateful for the archbishop's work in this regard, the Vatican was understandably cool toward the result.[17]

In the fall of 1902, Roosevelt appointed Storer as ambassador to Austria, and while the Storers were in Vienna, Mrs. Storer undertook the project to secure a cardinal's red hat for Archbishop Ireland. She sent her husband to the Vatican to ask the pope for this favor, implying that President Roosevelt wanted it, and in due course, word of these conversations reached the president's ears. Furious, Roosevelt withdrew Storer from Vienna so precipitately that the Austrians were offended. The State Department refused to say why Storer left Vienna, and for a time, nothing substantive was reported in the press; however, at the end of March 1906, Mrs. Storer gave an interview from Vienna. She quoted from a Roosevelt letter of 1900 in which he said, "I need not say what a pleasure it would be for me to do anything for Archbishop Ireland."[18]

Storer began working on his own version of his dismissal, which became a pamphlet dated November 1906 that Storer sent to the president and all members of the cabinet. In his pamphlet, Storer quoted a Roosevelt letter of the previous December, castigating Mrs. Storer for approaching the Vatican and asking her to promise in writing to leave church affairs alone. Then Storer flatly asserted that Roosevelt had asked the Storers to use their influence to secure Archbishop Ireland's promotion, and that he, Storer, had been dismissed only because Roosevelt's initiative on behalf of Ireland had become public knowledge.[19]

---

[17] Maria Longworth Storer asked McKinley to appoint her husband as first assistant secretary of state, but Sen. Joseph Foraker of Ohio, who did not like Storer, opposed the nomination. *New York Times,* 12 and 16 Dec. 1906. Storer was appointed minister to Belgium in 1897 and was transferred to Spain after diplomatic relations were restored with that country. *New York Times,* 19 March 1906. During the war, Filipinos occupied the so-called friars' lands of some 400,000 acres belonging to four religious orders, but the treaty with Spain guaranteed existing property rights. In order to avoid dispossessing the Filipinos, it was necessary to purchase these lands for $7 million. Negotiations with Pope Leo XIII were not productive, and the agreement was reached with the Vatican only after Leo died in July 1903 and Pius X was elected on 4 Aug. 1903.

[18] In March 1906, the *New York Times* speculated that Mrs. Storer's husband might have been removed because her nephew had married Alice Roosevelt. *New York Times,* 19 March 1906 and 1 April 1906.

[19] In the letter from Roosevelt, dated 11 Dec. 1905, there was reference to a critical letter from Cardinal Merry del Val, the Vatican secretary of state, accusing Mrs. Storer of being "unwarrantably officious." *New York Times,* 9 Dec. 1906.

Roosevelt then prepared his own public response, and the controversy exploded in the newspapers a week later. Roosevelt's reply appeared in the newspapers the day after Storer's defense of his action, and Roosevelt used quotations from correspondence to prove conclusively that the Storers had acted without his authority. Stung by this response, Mrs. Storer completed the public exchanges with an interview from Cincinnati on December 10, which was headlined "We Made Roosevelt" the next day in the *New York Times*. In it, she said, "It was through me and my influence that Mr. Roosevelt was made Assistant Secretary of the Navy." It was this supercharged topic that Plunkett discussed with Roosevelt at the dinner on December 18.[20]

A topic equally "indiscreet" and likely to irk the president was Plunkett's gratuitous criticism of the practice of distributing patronage geographically and also by religion. This subject probably arose in the conversation because one of the Roosevelt letters Maria Storer turned over to the press was one dated 1901, responding to Mrs. Storer's request for a cabinet post for her husband. In response, Roosevelt, who was casting about for excuses, noted that the spots Bellamy could fill were occupied, although he noted that he would like to have a Catholic in the cabinet. Then he also noted that Bellamy did not fit geographically (he was from Ohio), because he, Roosevelt, needed to bring in people from the Pacific Slope and from New England.[21]

Of course, no American president could ignore important sections of the country in selecting his advisors, and Plunkett was naïve to think that was possible. In the course of his criticism of the president's practice, Plunkett got a negative reaction from Oscar Straus, who thought Plunkett was referring to him, as he was Jewish. Straus, whose brother was co-owner of R. H. Macy & Co. in New York, was prominent in Jewish charities, and he had known Roosevelt in New York before the latter came to the White House. Straus was appointed minister to Turkey by President Cleveland and had

[20] Roosevelt's response was in the form of a letter dated 2 Dec. to Secretary of State Elihu Root. *New York Times,* 10 and 11 Dec. 1906. In January 1907, Plunkett called on Archbishop Ireland and discussed the Storer incident with him. Plunkett diaries, 3 Jan. 1907.

[21] The Roosevelt letter was dated 4 Oct. 1901 and was printed in the *New York Times* on 11 Dec. 1906. Also, Plunkett diaries, 18 Dec. 1906.

served twice in that office before becoming the first Jewish member of the cabinet. Withal, it was a remarkably indiscreet evening.[22]

Plunkett spent the next day at Johns Hopkins University, but two days later, he was back in Washington for another long interview with the president, Garfield, and Charles Patrick Neill, the U.S. commissioner of labor. Apparently, Neill was acting as the replacement for Straus, whom Plunkett had offended two days before, and there are no mentions of subsequent meetings with Straus. This meeting focused on Plunkett's rural life ideas, which had generated considerable interest by Roosevelt and Garfield, and Plunkett anticipated he would be asked to help prepare recommendations for the president. Later in the day, Dr. Frissell came to ask Plunkett to go to Richmond to discuss agricultural education with state officials.[23]

His Washington-area work done, Plunkett set off for the West to attend to his own business. On the train, he met Joe Cannon, the legendary Speaker of the House, who appears in the diary as "A man of the best Middle West type. Much of the Abe Lincoln pattern." The two men undoubtedly spent some time discussing Teddy Roosevelt. In Omaha, Plunkett learned that he still had partner problems, saying that Bosler was tricky, although not as bad as Windsor had been. He moaned that but for Windsor, he would be a millionaire (in dollars) but acknowledged that he now had more money than he needed, although not enough to completely free him for public life.[24]

After going over accounts with the auditor, Plunkett left Omaha

[22] When Straus was still minister to Turkey, he wrote a letter in the summer of 1904 abandoning the Democratic Party to support Roosevelt. *New York Times*, 20 July 1904. When Taft became president, Straus was returned to Turkey as ambassador in 1909. Although Roosevelt publicly denied that he wanted to appoint both a Catholic and a Jew to the cabinet, he told Straus he wanted American Jewish boys to have someone like Straus as a model. In 1905 he added Charles Bonaparte, a Catholic, to his cabinet. Cohen, *Dual Heritage*, 148.

[23] Plunkett diaries, 20 Dec. 1906. James Rudolph Garfield was a president's son who managed to have a fine career without suffering from his father's shadow. His father had been taking his son to college when he was assassinated, and the son witnessed the event. James Rudolph Garfield practiced law in Cleveland and served first on the Civil Service Commission then as commissioner of corporations in the Department of Commerce and Labor, before receiving appointment as secretary.

[24] Plunkett diaries, 19–22 Dec. 1906. Joseph Gurney Cannon was a regional prosecutor during the Lincoln administration. While Speaker of the House, Cannon made his much-quoted comment on Roosevelt, saying, "Roosevelt's all right, but he's got no more use for the Constitution than a tomcat has for a marriage license." Morris, *Rise of Theodore Roosevelt*, 11.

with Bosler to see the Diamond Ranch, where he spent two cold days. In Cheyenne he met former senator Joseph Carey and his wife, preparatory to visiting the Wheatland Project. The next day in Wheatland, Carey assembled the farmers to hear Plunkett talk about rural life, and Plunkett felt that if he had the time, he could organize these farmers more readily than he could his Irish farmers back home.[25]

In the new year, Plunkett again visited colleges to examine agricultural teaching, stopping at the Iowa college in Ames, then at the Minnesota college at St. Anthony, and finally in Madison, Wisconsin. Two more appointments in Chicago were followed by a visit to Hull House, where he met Jane Addams (a "fine woman"). A day later, he was at Cornell University in Ithaca, New York, meeting with Prof. Liberty Hyde Bailey, the first dean of the New York College of Agriculture. Plunkett was impressed with Bailey, and not just because the latter had read the essay he had prepared for President Roosevelt. Plunkett and Bailey later worked closely together on improving rural life in America.[26]

From Ithaca, Plunkett went back to Washington, D.C., to work with Pinchot and Garfield on agricultural policy initiatives for President Roosevelt. Plunkett found it easy to work with these men, whom he considered Roosevelt's "most intimate" advisors, and he now thought he was now doing the most important work of his life. "I must not be egotistical," he said, "but here was my life's thought & work being considered not in its application to Ireland, but to the . . . vast continent where the Irish race have made their home." Plunkett persuaded Pinchot to go to New York to meet Bailey and offer him the job of spearheading Roosevelt's rural life program. After a quick trip to Boston to visit Harvard and the Massachusetts Institute of Technology, Plunkett sailed for Queenstown, which he reached on January 25.[27]

Pinchot presented his work with Plunkett to Roosevelt and

---

[25] Plunkett finally regularized Conrad Young's full-time employment in the Omaha office at the beginning of 1907, agreeing to a salary of $3,000 per year. Plunkett diaries, 23–29 Dec. 1906 and 1 Jan. 1907.

[26] Plunkett diaries, 2–7 Jan. 1907. Liberty Hyde Bailey was born 15 March 1858 and in 1885 was a professor at Michigan State Agricultural College, where he established the first department of horticulture and landscape gardening in the United States. Three years later, he went to Cornell as chair of practical and experimental horticulture, the first such chair in the country. He traveled widely and produced over 700 titles on a broad spectrum of subjects.

[27] Plunkett diaries, 8–11 Jan. 1907.

afterward wrote to say that the president praised Horace in the presence of several people. Plunkett recorded the substance of Roosevelt's characterization of the Irishman in his diary in this fashion: "The only Englishman (!) he ever met whom he would like in his Cabinet & that if he could get me he wd put me in it in a minute." An important signal that Plunkett had shifted the emphasis of his work with governments from Great Britain to the United States was his formal request in April 1907 to be relieved of all public duties in the United Kingdom.[28]

Echoes still came to Plunkett from his ranching past, and in the summer of 1907, Boughton visited Plunkett twice. He had left the Ione ranch and the American West to rebuild his fortunes in Sumatra, Mexico, Singapore, and Australia, and now he had gray hair, was heavier, and suffered from a bit of rheumatism. He told Plunkett that he had often been "stoney broke" but always rebounded, and now had an income of £600–£700 per year, although he still had no settled occupation. He also told Plunkett that he envied the latter's public work. "This I did not expect," Plunkett said.[29]

Back in America again in the fall of 1907, Plunkett followed the pattern of the previous year, visiting universities and others to discuss his rural life ideas and making quick trips to look after his own investments. He went again to Harvard, this time to meet with Abbott Lawrence Lowell, professor of government. Then he went to New Hampshire to visit James Bryce, the new British ambassador to the United States, who had summer quarters in the White Mountains. From New Hampshire, Plunkett made a quick trip to Ottawa to speak with Canadian officials about rural life, but he did not record any details of this visit.[30]

The next stop was Omaha, where Plunkett spent time on his real estate investments and discussed his rural life views. He went to a public meeting on the annexation of South Omaha to Omaha, went to the university in Lincoln to discuss rural life, and in general familiarized himself with life in Nebraska. Then he went again to Madison, Wisconsin, and to the University of Illinois before going back to New York to discuss rural life at a dinner of millionaire benefactors. The thrust of the dinner was to secure money for

[28] Ibid., 24 March 1907 and 26 April 1907.

[29] Ibid., 6 June 1907 and 7 Aug. 1907.

[30] Ibid., 9–16 Oct. 1907. Abbott Lawrence Lowell began teaching at Harvard in 1897 and became president of Harvard in 1909. James Bryce, who had been chief secretary for Ireland, was named ambassador to the United States in February 1907.

his work in Ireland, and Plunkett was hopeful of help from Rockefeller.[31]

In Washington, D.C., Plunkett met with Pinchot and Neill then, after a trip to the Hampton Institute, came back to stay at Pinchot's house. The next day he lunched with the president again, and Mrs. Roosevelt arranged for him to meet her governess, who was Conrad Young's older sister. In the afternoon, Roosevelt showed him the references to agriculture in his annual message to Congress: "practically my own suggestions," Horace wrote. At Roosevelt's request, Plunkett and Pinchot retired to prepare a memo for the president to send to Secretary of Agriculture Wilson, suggesting the establishment of a Bureau of Rural Social Emigration. The idea for the bureau was Plunkett's, and he was already contemplating who should run it.[32]

Unfortunately, Roosevelt was always the politician, and Plunkett soon learned that Roosevelt had not explained Plunkett's ideas to old James Wilson. Instead, Roosevelt left this chore to Plunkett, thus retaining the option to withdraw gracefully if Wilson's reaction were too adverse. After Plunkett and Pinchot met with Wilson, Plunkett conceded that "my Bureau still hangs in the balance," but he was hopeful. Although Roosevelt believed in Plunkett's rural life ideas, his increasingly difficult relationship with Congress made it impossible for him to accomplish everything he wanted to do before the end of his administration. The rural policy agenda was still a piece of unfinished business when he handed over the White House to Taft.[33]

Early in 1908, another chapter was written in the long story of Moreton Frewen's troubles with Plunkett. Frewen's other creditors were suing to recover what he owed them, and Horace, who was also a creditor, wrote to Moreton to say he would not take part in that action. "My claim against you is [just]," Plunkett wrote, adding, "Should you ever as I hope you may, be in a position of affluence you will then be free to consider whether it should be revived at your instance." Although no money changed hands

---

[31] Ibid., 3–19 Nov. 1907.

[32] Plunkett thought Gertrude Young, who was born in England in October 1862, was superior to her brother in culture and mental ability. Plunkett diaries, 22–25 Nov. 1907.

[33] Secretary Wilson, who was born in 1835, had been secretary of agriculture under McKinley, and he remained in the position through the Roosevelt and Taft administrations. Ibid., 27 Nov. 1907.

because of this letter, the two men resumed writing to each other, albeit not about cattle but about Ireland and politics.[34]

In the waning months of his administration, Roosevelt pressed on with his ideas for improving rural life in America, and to get around the increasingly fractious Congress, he used a concept he had used before: the presidential commission. Other presidents had used commissions to deal with special situations, but Roosevelt made them into an instrument of executive policy, beginning with his first commission in 1902 and reaching a total of eleven by the time he left office. At the beginning of August 1908, Roosevelt appointed his Commission on Country Life, and following the advice of Pinchot and Plunkett, he appointed Professor Liberty Hyde Bailey as the chairman.[35]

After appointing the commission, Roosevelt asked Plunkett to call on him the moment he got to the United States. Plunkett did call on the president, but first he made a brief side trip to Omaha to learn more about the commission. The Country Life Commission was sitting in Omaha at the time, and Plunkett met Chairman Bailey and Henry Wallace, of *Wallace's Farmer,* a commission member. In Washington, D.C., Plunkett again met with Roosevelt, who expansively told the others present that Plunkett had an "international position" in connection with the Country Life Commission idea. Afterward, Plunkett and Pinchot worked on the commission's report and the accompanying message for Roosevelt to send to Congress.[36]

---

[34] Horace Plunkett to Moreton Frewen, 25 Jan. 1908, Frewen collection. Plunkett outlived Frewen by eight years, and in the year before Plunkett died, Clara Frewen's nephew, Shane Leslie, came to see Plunkett about writing a biography of Moreton. Plunkett thought their conversation disillusioned the potential author, and indeed, Leslie never wrote the book—whether for that reason or some other, we do not know. Long after Plunkett's death, Leslie's daughter, Anita Leslie, wrote *Mr. Frewen of England: A Victorian Adventurer,* based in part on her father's research. Also, Plunkett diaries, 11 Feb. 1931.

[35] Page Smith called Roosevelt's use of commissions "a bold extension of executive authority." Page Smith, *America Enters the World,* 126. The original members of the commission, in addition to Bailey, were Henry Wallace, editor of *Wallace's Farmer;* President Kenyon L. Butterfield of Massachusetts Agricultural College; Gifford Pinchot; and Walter Hines Page, editor of *The World's Work* and future ambassador to Great Britain. In November 1908, Roosevelt added Charles S. Barrett, president of the Farmers Cooperative & Educational Union of America, and William A. Beard of *Great Western Magazine* to the commission. *New York Times,* 10 Aug. 1908 and 16 Nov. 1908.

[36] When Plunkett later went to Wheatland, he also was on an errand for the Country Life Commission, although he did not give details of that work. Plunkett diaries, 23 Sept. 1908; 9, 10, 11, and 19 Dec. 1908.

Two days later, Roosevelt met Plunkett again and once more stressed how important Plunkett's help had been to the rural policy. Then, as Plunkett had done with Secretary Wilson, Roosevelt asked Plunkett to explain the rural policy to Taft, who had just won the White House in the fall elections. While Plunkett waited, Roosevelt dictated a letter to Taft, lauding Horace "to the skies" and asking the president-elect to see him. If either Plunkett or the president saw the anomaly in asking a foreign national to brief the newly elected president on a key matter of domestic agricultural policy, no word of it reached Plunkett's diary.[37]

Taking the president's letter with him, Plunkett rushed down to Augusta, Georgia, where Taft was vacationing. Plunkett found that Taft was totally ignorant of Roosevelt's rural policies in both principle and detail. In the time available, Plunkett did his best to explain the policy and emphasized that Taft's secretary of agriculture must have sociological sensitivity. Mindful of his criticism of Roosevelt's patronage policies, Plunkett worried—at least in his diary—that the need for geographical distribution of patronage would inhibit the choice of the right man for the job.

Explaining all this to Taft took twice the time Plunkett allotted for the visit, and Taft asked if he could not stay another day. Unfortunately, Plunkett had to meet with Roosevelt again on Christmas Eve and had to leave immediately. When Plunkett left, the 320-pound Taft slapped the 130-pound Plunkett on the back, much as Roosevelt would have, which must have been an interesting sight. Taft said, "Tell the President he may count me one on his policy—I am not quite sure that I understand it all, but you and Gifford Pinchot must see me through!" Taft then sent away the carriage and took Plunkett to the station in his own automobile. While Plunkett's reception by Taft was very gratifying to him, he later learned that like Roosevelt, Taft was also a good politician.[38]

Plunkett saw Roosevelt again in Washington, where Plunkett reported on his interview with Taft and the two men also discussed many other topics. One of those other topics Roosevelt mentioned

---

[37] Plunkett diaries, 21 Dec. 1908.

[38] Plunkett's detailed notes on his visits with Taft and Roosevelt were contained in his letter of 31 Dec. 1908 to Lady Betty Balfour, quoted in Digby, *Horace Plunkett,* 127–30. Lady Elizabeth Edith Bulwer-Lytton, daughter of Lord Lytton, married Gerald William Balfour, who had been an Etonian classmate of Plunkett and who served as chief secretary of Ireland for five years. Plunkett used Lady Betty as conduit for information he wished to pass on to Balfour.

was the storm caused by remarks he made earlier in 1908 about the budget for the Secret Service. Reacting to congressional criticism of his appropriation request for the Secret Service, Roosevelt said members of Congress were afraid they would be investigated. Roosevelt told Plunkett he regretted his choice of words. This comment gave Horace the opportunity to give the president a lesson on the English language. He even urged Roosevelt to avoid split infinitives, reminding the president that college professors read his writings. Roosevelt promised never to let another slip by. After leaving the president, Plunkett worked on Roosevelt's message to Congress, happily noting in his diary, "Making history." He sailed for home at the end of 1908, comforted by a dream of making Plunkett House in Ireland the research laboratory for the rehabilitation of country life in the western world.[39]

Roosevelt sent the report of the Country Life Commission to Congress early in February 1909 before he left office, recommending that the Department of Agriculture be reorganized into a Department of Country Life. He also recommended increased powers for the department, so that it could have more influence over the quality of rural life. This beginning was very encouraging to Plunkett, but unfortunately, the work of the commission quickly ran into opposition.[40]

The report of the commission recommended comprehensive surveys of the agricultural regions and a redirection of public education in rural areas. The report emphasized the importance of working together and lifted up the country church as an instrument for moral improvement. The recommendations ended with the call for training young leaders for the rural areas. Some sections of the report clearly reflect the specific interests of its authors: the sections on wastage and control of forests and streams bear Pinchot's imprint, and the recommendations on working together are, of course, from Plunkett.[41]

A good portion of congressional opposition to the Country Life Commission's report was merely a boiling over of frustration from

[39] Plunkett called his concept a Bureau of Rural Social Economy for the Western World. Plunkett diaries, 23 and 26 Dec. 1908 and summary for the year 1908.

[40] *New York Times,* 10 Feb. 1909.

[41] *Report of the Commission on Country Life.* The report was originally published for the use of Congress as *Senate Document No. 705, 60th Congress, 2d Session* but was not released to the public, and only the first thirty-one pages of the 150-page report were given to the press.

past failed confrontations with the activist president regarding his commissions. Even though Country Life Commission members were unpaid and their expenses were absorbed by the Russell Sage Foundation, congressmen just wanted to kill all of the Roosevelt commissions and their reports.

Outside of Congress, James J. Hill also opposed the Country Life Commission. Hill had his own reasons to be unhappy with Roosevelt's ideas after having been forced by the Supreme Court to dismantle the Northern Securities Company, but he also had a great interest in scientific farming. Although the commission held thirty public hearings—which drew farmers from forty states—and obtained answers to 130,000 questionnaires, Hill contended that the commission did not obtain enough information. Hill also argued that Plunkett's program for better education would not produce results quickly enough, saying, "It will not do to wait until the children grow up, go to the agricultural college and come back to the farm, fifteen or eighteen years hence."[42]

When the Senate considered the president's message forwarding the report of the Country Life Commission, the message was greeted with "open amusement." The conference committee considered the appropriation for $25,000 to publish the commission's report, but the House refused to approve it. Congress made its sentiment clear in a rider to the Sundry Civil Appropriations Bill, which forbade the creation of any more commissions without congressional authority.[43]

The objectives outlined by the Country Life Commission were noble, and the report was the first effort to establish social work in rural areas in the United States. Although it has been said that the report "probably had more influence on rural life in this country than any other document," the recommendations did not produce results for a long time. Early on, nothing was done, and the agri-

[42] Hill, Morgan, and Harriman formed the Northern Securities Company in 1901, and the next year Roosevelt ordered the attorney general to bring suit against it, which resulted in the breakup of the company in the spring of 1904. Hill also criticized Pinchot's actions in reserving natural resources from use by the public and called for "economic, scientific and self-perpetuating *present use*" of them (emphasis added). Martin, *James J. Hill*, 550.

[43] Roosevelt enraged many in Congress when he added sixteen million acres to forest reserves just before he signed a bill that forbade him to do just that, and Pinchot made more congressional enemies by designating 2,500 sites as ranger stations to keep them from being used as water power locations. The Spokane Chamber of Commerce ultimately subsidized printing of the Country Life Commission's report. Pinchot, *Breaking New Ground*, 341, 343.

cultural depression of the twenties took its toll before there could be any real progress in rural social work.[44]

Roosevelt was anxious to give some public recognition to Plunkett for his work with the U.S. government, and just before he left the White House in March of 1909, he wrote to the British ambassador. His letter said, in part, "We Americans owe much to Ireland and Sir Horace Plunkett in the work we are keen in trying to do in the United States." Unfortunately, when the letter reached London, Plunkett's enemies there found reasons to keep it secret.[45]

When Plunkett came back to America in the fall of 1909, he made the usual trip to the Diamond Ranch, where he found the company was badly in need of marketing skills to sell its land—skills not available inside the company. The shareholders tried to sell the entire company, but there were no offers, and it was now obvious they would have to undertake the laborious process of making piecemeal sales of land. In the meantime, of course, the company needed more money.[46]

Plunkett turned next to the Wheatland Project, where another management problem loomed. Joseph M. Carey was 65, and time had taken its toll on him, for he had been nearly deaf as early as 1896, making it difficult for him to continue as the chief executive of the company. To deal with this problem, Plunkett pressed Carey to give his son Robert a share of management duties. This effort bore fruit, and Robert took responsibility for the two Wheatland companies at the end of 1910. Joseph Carey's trust in Plunkett was further exemplified by the fact that it was Horace, acting for the board of directors, who formally handed over the new responsibilities to Robert.[47]

---

[44] The quotation is from Dwight Sanderson, in Olaf F. Larson & Jones, "Unpublished Data from Roosevelt's Commission," 583. This article notes that the 94,000 responses to the circular questions sent out by the commission are a trove of information about the condition of country life in the United States. Also, Martinez-Brawley, *Seven Decades of Rural Social Work*, 1 and *passim.*

[45] The letter to Ambassador James Bryce was dated 2 March 1909. *New York Times,* 29 May 1909. The mean-spirited effort by the British government was foiled by Pinchot, who had the letter published in the United States in *Wallace's Farmer* so that it could then be reprinted in Ireland in the *Homestead,* which was the organ of Plunkett's Irish Agricultural Organisation Society, on May 28. From thence, the handling of the letter was reported in other British newspapers, and eventually questions were raised in Parliament, making for a few days' lively controversy.

[46] Horace Plunkett to Frank C. Bosler, 4 Nov. 1909 and 29 April 1910, Bosler collection.

[47] Plunkett diaries, 1 and 2 Nov. 1909 and 19 Dec. 1910. Judge Carey was absent, back in the East, when Plunkett arrived in Cheyenne in December 1910.

Back in New York, Plunkett turned to the problem of raising money for his Irish work, but a three-day effort with Andrew Carnegie produced no results. He then turned his attention to John D. Rockefeller, Jr., at a dinner at the Rockefeller house on 54th Street. The house was a rented four-story mansion (Plunkett called it a "small pokey house") across the street from the sumptuous brownstone of John D. Rockefeller, Sr. Rockefeller, Jr., could be forgiven for lack of attention to his city residence, as he was then deeply involved building Kykuit, the Rockefeller mansion on a 3,000-acre estate in the Pocantico Hills of Westchester County, New York. Kykuit would definitely not be a "pokey" house.[48]

In Norfolk, Virginia, Plunkett heard President Taft speak, and afterward Taft invited him to an oyster roast. But the old magic Plunkett had enjoyed with Roosevelt was gone, and Taft was already replacing Plunkett's friends in government. At the Interior Department, Garfield was already gone, and his successor tried to reverse Roosevelt's protection of almost a million acres of forest lands. Pinchot brought this matter to Taft's attention, and when he was not satisfied with the result, he went public in *Collier's Weekly.* (Plunkett thought Pinchot's action was "not quite proper.") Pinchot expected to be dismissed, and Plunkett worked with him to draft a letter for public consumption, in expectation of that action, which came in January 1910.[49]

Thus ended Plunkett's first efforts to make history in the United States, cut short by the departure of Roosevelt in the spring of 1909. Although he later moved in official U.S. government circles again, he never found an American president so receptive to his ideas.

---

[48] John D. Rockefeller, Sr., lived at No. 4 West 54th Street, in the four-story mansion of Collis Huntington, and three other Standard Oil directors lived nearby. Chernow, *Titan*, 221–22, 359, 402. John D. Rockefeller, Jr., lived at No. 13 West 54th Street until 1913, when his house at No. 10 West 54th Street was completed. Plunkett wrote of Rockefeller, Jr., "A narrow guaged good religious dull man. No signs of wealth in his message though he married a daughter of Sen. Aldrich (also *nouveau riche*). No wine—small pokey house (13 W. 54th) second rate cooking, 3 rather dingy man-servants, dull company, exc. Dr. Flexner, head of Rockefeller Inst. For Medical Research." Plunkett diaries, 18–21 Nov. 1909 and 21 Dec. 1909.

[49] Plunkett saw Pinchot and Garfield in New York after he landed in October, and they talked, contrasting the good days of Roosevelt with the "evil days" of the "judicial" Taft. Plunkett diaries, 19 Oct. 1909, 25 Nov. 1909, and 10 Jan. 1910. Pinchot had made many enemies in his zealous pursuit of conservation, and he was soon engaged in a controversy with the new secretary of interior, Richard Achilles Ballinger.

11

Writing and Convalescing

In 1909, Plunkett began to consider what effect his death might have on his partners in American investments, and he set about to buy up these interests. He borrowed the money to pay off his partners, and the interest on this debt, together with a voluntary payment to his old Frontier Company partners, used up his real estate gains of more than £5,000 in 1911.[1]

A new element was added to Plunkett's American travels in the fall of 1909: a search for help for his declining health. In Des Moines, Iowa, he visited Still College, the second oldest osteopathic medical school in the United States. After diagnosis, he submitted to treatment, which he repeated twice with other osteopaths. During this period, he also devoted more time to his writing.[2]

As noted earlier, Plunkett's first book dealt with Ireland, but his next book was an outgrowth of his work with Roosevelt on the rural life problem in the United States. Dr. Lyman Abbott, editor of *The Outlook,* got Plunkett to write a series of articles on the subject, which he began working on at the end of 1909; he sent the fourth, and last of this series, to the magazine in January 1910.

---

[1] Plunkett diaries, 26 Oct. 1909, 19 Jan. 1910, and 2 March 1910. Not all the details of the buyout offers are known, but one was made to Lord Monteagle; £8,000 was offered to Freddie Blacker, the heir to Willie Blacker's interests; and a similar arrangement was made with Alston. Thomas Leonard, another of Plunkett's partners in real estate, accepted a buyout of his £1,000 investment at the end of 1892. Thomas Leonard to Horace Plunkett, 12 Dec. 1892, Plunkett collection, Library of Congress.

[2] The diagnosis Plunkett received was "general neurosis occasioned by increased dorsal curve with extreme rigidity in the low cervical and dorsal area of the spine." Plunkett diaries, 10, 12, and 14–15 Nov. 1909.

After the *Outlook* articles were published, Plunkett converted the four articles to a book of seven chapters, *The Rural Life Problem of the United States: Notes of an Irish Observer.*[3]

In the first article, he focused on the need for conservation by farmers, declaring that the farmer, "now the chief waster," must become the chief conservator. Achieving this transformation would involve both social and political change, but he expected that economic methods would be the chief agent of these changes. In the second article, Plunkett called attention to the movement of people from farms to urban centers, which he likened to the depopulation of rural districts in England and Ireland that, in turn, exacerbated the farm labor shortage. He thought this movement stemmed from the decline of individual ownership of farms and the growth of tenancy.[4]

In the third article, Plunkett dealt with public opinion bias in favor of urban life. Defending himself against complaints that his emphasis on rural life was outdated, he pointed to the speeches at the annual Farmer's Club dinner at Delmonico's in New York City as evidence of current interest in the subject. (Whether the state of the American farmer could be perceived at a farmer's dinner in New York City was at least questionable.) Moreover, when he used a reference to Virgil's handbook on Italian agriculture in making his argument, he must have puzzled his American audience.

He introduced his American audience to the "better farming, better business, better living" formula for improving the quality of rural life that he had already used in Ireland. Farmers could achieve all these objectives by working together in cooperative associations. Nevertheless, Plunkett recognized that independent-minded American farmers could not easily be organized, and he especially questioned whether black farmers in the South would join the movement.[5]

In the final article of the series, Plunkett advocated better agricultural education for farmers, because fewer than 1 percent of the Midwest farmers had any contact with agricultural colleges and universities. He contended that cooperatives could also improve communications between farmers and educational institutions.

[3] Plunkett diaries, 20 Nov. 1907, 7 Dec. 1909, and 31 Jan. 1910.
[4] Plunkett, "Conservation and Rural Life" and "Human Factor in Rural Life."
[5] Plunkett, "Better Farming, Better Business, Better Living."

Finally, he asserted that cooperatives could enhance the role of the woman in rural life by easing her burdens.[6]

The Rural Life Problem of the United States: Notes of an Irish Observer appeared in July to generally friendly reviews. One reviewer said, "There is thought in every line. The gist of the case for the reconstruction of country life has never been better condensed." The *New York Times* called the book "meaty," and said the work was able, informed, and constructive, although the reviewer didn't like the title. Distribution of the little book was considerably aided by Lord Grey, governor general of Canada, who sent 400 copies to western Canada.[7]

Plunkett came to the United States in December 1910 to check on his remaining U.S. investments and to seek medical treatment. He stopped in Omaha, where he increased Conrad Young's compensation, and then went to Cheyenne to meet Robert Carey, who was now in charge of the Wyoming Development Company. While in Cheyenne, he discussed the Wheatland Project, the Diamond Ranch, and some remnants of the Frontier Company. A new writing opportunity appeared when Plunkett learned that the elder Carey, who was now the Wyoming governor, needed to deliver an address to the legislature, and Plunkett volunteered to help with the task.[8]

With business matters out of the way, Plunkett turned to the serious matter of his health. At the end of December, his friend Gifford Pinchot, who was preparing to enter the Battle Creek Sanitarium, urged Horace to go there as well. It was in this fashion that Plunkett began his long association with the sanitarium at Battle Creek, Michigan, managed by Dr. John H. Kellogg. Plunkett's many visits to that institution were the source of great physical comfort to him and may well have prolonged his life, which was troubled by physical weakness and disease.[9]

Plunkett grew up in an era when medical practices were primi-

---

[6] Plunkett, "Better Farming, Better Business, Better Living: Two Practical Suggestions"

[7] Plunkett diaries, 18 and 21 Feb. 1910, 9 April 1910, 21 July 1910, and 5 Jan. 1911. Albert H. G. Grey, 4th Earl Grey, served as governor general of Canada from 1904 to 1911. He is not to be confused with Sir Edward Grey (later Viscount Grey), who was British foreign secretary. Also, *New York Times,* 20 Aug. 1910.

[8] Plunkett paid Young 10 percent of profits and 5 percent of property sales. Plunkett diaries, 8 and 19–24 Dec. 1910.

[9] Ibid., 24 Dec. 1910.

tive, at best, and the Plunkett family devoted a fair amount of study to the subject. His father decreed a "spring and autumn cleaning" for the bodies of his children. This formidable regimen consisted of heavy doses of calomel, mixed with raspberry jam, followed the next morning by either castor oil or the laxatives: Glauber salts or Gregory's Powder. Later, there were occasions when Horace and his brother even felt confident enough to challenge the doctor's diagnosis of their father's illnesses. As in other matters that attracted his interest, Horace eagerly researched the subject whenever he had the chance.[10]

For a number of years, particularly when he was ranching, Plunkett suffered from persistent diarrhea, the outbreak of which correlated well with the stress in his work. He tried, without much success, to alleviate the problem with diet. Nor did his health improve after he left the strenuous life on the western cattle ranges, for long periods of concentrating on writing would also disable him. His physical troubles increased when he broke his leg in an accident and the break did not heal, and while he was still convalescing from the first fall, a second fall reopened the fracture. The medical use of X-rays was a recent discovery, and Plunkett's first contact with this new technology occurred when his fractures were treated. Later, overuse of X-rays gave him a nasty burn that required surgery.

Plunkett arrived at the Battle Creek Sanitarium at the end of 1910, eager to study the Kellogg system of treatment. The Battle Creek Sanitarium was owned by the Seventh Day Adventists, who opened a hydrotherapy clinic in 1866. The church sent John Harvey Kellogg to a hydropathy college in 1872 and helped to finance his medical studies at New York's Bellevue Hospital, which was one of the better medical schools. Kellogg graduated from Bellevue in 1875, and the next year he was named medical director of what was then a failing institution. Dr. Kellogg was an enthusiastic advocate

[10] Ibid. Although there are many entries regarding medicine in the Plunkett's diaries, it is interesting that he claimed the only medical work he had read was written by the court physician to Charles II. A medical work of Sir Isaac Newton's time prescribed dried wolf's intestines and sheep's excrement as a treatment for colic. Also, Plunkett, *Some Tendencies of Modern Medicine*, 24, 33. Glauber's salt, also known as *sal mirabilis,* was a laxative discovered by Johann Rudolf Glauber (1604–1670). Gregory's Powder, invented by Professor James Gregory (1753–1821), consisted of pulverized rhubarb, ginger, and magnesia.

of vegetarianism, and although at 5'4" he was not am imposing figure, he soon dominated the Battle Creek establishment.[11]

Kellogg wanted the sanitarium to be a place where people could learn how to stay well, and he was also a born promoter who catered to famous patients. Horace Plunkett was one of that favored group, which included William Jennings Bryan and former president William Howard Taft, who was celebrated as the 100,000th patient. During Plunkett's periodic visits to Battle Creek, Dr. Kellogg generally prevailed on him to address the other patients and sometimes people from the town, as well. The subjects of these talks were generally topical, and more than once, Plunkett spoke of conditions in Ireland. Kellogg had his detractors, but his methods succeeded so well financially that he was able to wipe out the debt, and at least those critics inside the organization were silenced. Not incidentally, his work with vegetarian diets eventually made it possible for his brother, Will K. Kellogg, to launch a very successful cereal business.[12]

At Battle Creek, Dr. Kellogg took Pinchot and Plunkett under his especial care. The first day began with a "test meal," a breakfast of bread and water that was pumped out of the stomach after an hour to see what the stomach was doing with it. Massage, gymnastics, and rest occupied the entire day. The presence of reporters was an integral part of the schedule. They came chiefly to talk to Pinchot, relegating Plunkett to "a second string," according to his diary.[13]

The next day, Plunkett swallowed a concoction of sour milk and bismuth, and then he was "radiographed" to show dilated and con-

---

[11] Kellogg originated the name Battle Creek Sanitarium using a variant of the word "sanitorium." Despite his assertion that the sanitarium was "undenominational," Kellogg did recognize Saturday as a day of rest, on which no procedures were conducted. Although the Battle Creek Sanitarium later opened a branch in the United Kingdom, Plunkett continued to make his regular pilgrimage across the ocean to see Dr. Kellogg, himself. In addition to bad diet, Kellogg was fixated on sexual excess as a major enemy of good health. Soon after he became medical superintendent of the Battle Creek Sanitarium, he wrote *Plain Facts for Old and Young,* a book of over 500 pages ranging widely over all aspects of sexuality, including hints to detect excess in the young (round shoulders was one). Several editions of the book were published beginning before 1879.

[12] Dr. Kellogg was born 26 Feb. 1852 and died 14 Dec. 1943. Plunkett diaries, 23 Dec. 1915; Schwarz, *John Harvey Kellogg,* passim; and Sokolow, *Eros and Modernization,* 162. Dr. Kellogg's younger brother, Will Keith Kellogg, was in charge of producing cereal for former patients of the sanitarium. After C. W. Post launched a cereal using a flaking process similar to the one used by Kellogg, Will K. Kellogg left his brother's sanitarium to create his own company in 1906.

[13] Plunkett diaries, 29 Dec. 1910.

tracted organs. Dr. Kellogg believed that nearly all medical troubles resulted from poisoning by the wrong diet, and he took Plunkett off coffee, alcohol, and tobacco. According to the doctors, Plunkett's radiograph showed that his stomach was dilated and extended too far downward by three inches, and his colon was also constricted. The services of the sanitarium also included a dentist. In due course, Plunkett learned how to grind his cereal properly after he went home.[14]

Saturday at the sanitarium was observed as the Sabbath, so on that day most of the staff were gone; to pass the time, Pinchot gave a lecture on conservation to the "inmates" and people from town. A few days later, Plunkett again spoke on an Irishman's thoughts on conservation. Plunkett spent two weeks at the sanitarium and left for New York on January 15. By the end of February, he had still not received a bill from Kellogg, and on checking he learned he had been a guest at the institution.[15]

In New York Plunkett ran into Henry James, the novelist, and took advantage of the opportunity to ask James for his view on American civilization. James replied that he was disgusted with its crudity and unmannerliness and indicated he was wedded to the New England tradition. Plunkett tried in vain to get him to see that this tradition was being "snowed under" by "the hordes of barbarians" from Eastern Europe.[16]

Near the end of February, Plunkett went to the University of Wisconsin at Madison, which he had visited twice before in 1907. He may have met Dr. Charles McCarthy before, but this time he specifically mentioned their discussion. McCarthy was then 37, a decade younger than Plunkett, and was the head of the Legislative Reference Library at Madison, a title that badly understated the importance of his influence on social legislation in Wisconsin. McCarthy helped farmers to secure helpful legislation, and as a member of the State Board of Public Affairs he worked to advance a broad program of agricultural aid. When Plunkett met him, McCarthy was about to

[14] Ibid., 30 Dec. 1910.

[15] Plunkett contributed $150 to the sanitarium. Plunkett diaries, 31 Dec. 1910–14 Jan. 1911 and 28 Feb. 1911. Dr. Kellogg did not take a fee for his own services (including surgery) at the sanitarium, but it appears that on this occasion Plunkett was also excused from any other charges, as well.

[16] Ibid., 15–16 Jan. 1911. Plunkett may have met James before, as the latter was a friend of the London Sturgises.

publish *The Wisconsin Idea*, in which he expounded his thesis that monopolistic forces were increasing poverty in the United States. McCarthy was later a pivotal force in carrying out Plunkett's ideas for agricultural cooperation in the United States.[17]

By the tenth of March, Plunkett was back in England, and later in the spring, he wound up the last remnants of the Frontier Company. With the transfer of the northern herds to the American Cattle Trust, the Frontier Company had no active operations, but its affairs were not finally wound up for some years. The company owed $60,000 at the time of the transfer to the trust, of which one-third was owed to banks, one-third to Plunkett, and one-third to another unnamed shareholder (probably Boughton). Plunkett agreed to be responsible for the bank borrowings, and the shareholders then agreed to use the proceeds from the American Cattle Trust (and its successor companies) to pay off the debts, distributing any surplus to the shareholders.[18]

John Chaplin was responsible for collecting and disbursing the proceeds, and the first distribution—made in 1887—was used to reduce debt. By 1894, the shares of the trust's successor company were practically worthless, and Chaplin asked the shareholders to bid for them. Plunkett submitted the only bid, paid off the other creditor, and in the years that followed he realized a profit on his investment in the shares, which he later shared with his old partners. This voluntary sharing does not mean Plunkett had recovered all of his own investment in the Frontier Company, although indications are that he fared better than most of the British cattlemen in that disastrous period for the range cattle industry.[19]

---

[17] Plunkett visited the University of Wisconsin twice in 1907 but apparently only visited the agriculture college on those occasions. McCarthy visited Denmark to learn of their cooperatives and studied other examples in Australia and New Zealand, but he always freely credited Plunkett for his contribution to the beginnings of the American cooperative movement. Plunkett raised the first money for the American Agricultural Organization Society. McCarthy, *Wisconsin Idea*, 8; and Fitzpatrick, *McCarthy of Wisconsin*, 175–76.

[18] Plunkett diaries, 10 March 1911.

[19] In a letter to Chaplin, Plunkett explained that his distribution was voluntary, so that he could decide who should receive payment. The Roche brothers; Booth; Willie Blacker's heir, Maxwell; and Chaplin were given a pro-rata distribution. The unnamed others who had not been active in the business received nothing. Finally, Boughton was also excluded from final payment, because he had been a large shareholder but still elected not to share the final risks with Plunkett. One unnamed shareholder was undoubtedly V. G. Lantry, who had sold the Herman property to the Frontier Company. Other shareholders not mentioned in the list were John Coble and Plunkett's Eton schoolmate William Douglas Watson-Smythe.

An unexpected potential threat to the Wheatland Project arose down in Colorado, where a ditch diverting water from the Laramie River threatened the water rights for the big storage facilities of the Wheatland Project in Wyoming. The so-called Skyline Ditch diverted water from the Laramie River to irrigate lands in the drainage of the Cache la Poudre River in Larimer County, Colorado. Unfortunately, this diversion, which had been authorized by Colorado authorities, commenced in 1890 and was earlier than some of the Wheatland Project water rights. The State of Wyoming contested this diversion in 1911, but the U.S. Supreme Court ultimately upheld the Skyline appropriation. The Court limited the Colorado withdrawals from the Laramie River, and the Wheatland Project was not seriously damaged.[20]

Despite the session at Battle Creek at the beginning of the year, Plunkett's health still bothered him, and unfortunately he received conflicting medical advice from several sources. The Dublin doctors told him that his head was "wasting" his body, whereas Plunkett quoted Dr. Kellogg as saying that "my digestive apparatus is playing the devil with my nerves." He was so light-headed he could not attend the coronation of George V in Westminster Abbey, although he had a seat, and he checked into a nursing home in England. The English doctors told him not to go to Ireland before fall, and he spent some of his time writing a long argument against the doctors' diagnosis. In August, a new doctor was no help either, as he sighed that he couldn't "penetrate the professional crust."[21]

At the end of 1911, Plunkett determined to seek help again at Battle Creek, and he arrived there the day after Christmas only to learn that Dr. Kellogg was in Europe. Dr. Kellogg returned on January 19 and told Horace he needed eight weeks to "fatten." Plunkett undertook treatment but interspersed it with other activities away from Battle Creek. He made a quick trip to Chicago for a meeting, returned to Battle Creek for two days of X-rays, addressed 1,200 students at the Lansing agricultural college, and did a great deal of writing for the Irish newspapers. In the sanitarium, Plunkett spoke

---

[20] The Laramie River dispute was resolved in the U.S. Supreme Court in *The State of Wyoming vs. the State of Colorado, the Greeley–Poudre Irrigation District, and the Laramie–Poudre Reservoirs & Irrigation Company*, 309 U.S. 572 (1922).

[21] Plunkett diaries, 22 May 1911, 3–19 May 1911, and 4 Aug. 1911.

on "Ireland Today," which emptied the room of all except about 50 hardy souls. Finally, Dr. Kellogg gave reluctant permission for Plunkett to leave the sanitarium on March 6 (the end of ten weeks), directing him to have a nurse with him for six months.[22]

After leaving Battle Creek, Plunkett went to Chicago to meet with Johnny Peirce, Conrad Young, and Bosler to talk about Diamond Company affairs. Unfortunately, what he heard at this meeting and subsequently in the following year was a source of great frustration to him. Acting without consulting Plunkett, Bosler had committed a great deal of money to build a large irrigation project on the Laramie Plains. Plunkett was then forced to advance his share—and more—in order to protect his investment. Bosler wrote long letters explaining his reasoning, always after the fact, and he never apologized for failing to consult Plunkett in advance. Instead, he emphasized that he had advanced even more money than Plunkett did, which was true, but Plunkett always seemed to be the one who had to contribute the critical dollars to keep the project from shutting down.

The irrigation project was carried out by the Rock Creek Conservation Company, a wholly owned subsidiary of the ranch company, while the parent company continued to own and operate the cattle business, with Johnny Peirce as foreman. Some of the land to be developed was owned in fee by the parent company, whereas other land was public domain, where the settlers would buy water rights from the Rock Creek Company and file on the land with the government. The parent company provided the development capital, and the Rock Creek Company was to repay the parent company's advances from its sales of land and water to settlers. Income from the reclamation work was to be shared between the two companies, with the Rock Creek Company receiving income from the sale of water rights and the parent company receiving income from the sale of its patented land.[23]

Unfortunately, the market was crowded with a number of competing development schemes in Wyoming. Plunkett, himself, was

[22] Ibid., 26 Dec. 1911—12 and 17–28 Feb. 1912.

[23] The Rock Creek Water Users Association administered the water rights sold to settlers. Frank C. Bosler to Horace Plunkett, 9 March 1912, Bosler collection. The Diamond Co. project also made filings under the Carey Act, pursuant to which 4,400 acres were segregated and about 4,300 acres were ultimately patented.

involved in the large Wheatland Project, which had considerable land for sale, and the northern part of the Wind River Reservation had just been opened to settlement, creating another large inventory of potential farmlands. Up in the Big Horn Basin, there were also several other large reclamation projects. Finally, one of the most troubling potential competitors was a 40,000-acre tract northeast of Cheyenne, where a group promoted a dry farming venture that did not require expensive irrigation works.[24]

Once again, Plunkett liked the concept of the project, saying he could use the profits to finance his work in Ireland. However, he emphasized to Bosler that he wanted to limit his obligations to his stock interest, or a little over 15 percent. Bosler could not raise all the money he needed but pushed forward with the work anyway, using short-term borrowings to pay the bills. The ditches were all completed in time for the 1911 farming season, but by the middle of the year, total expenditures on the project were over $305,000, which was $100,000 more than original estimates.

To generate revenue until settlers took up the land and to demonstrate the feasibility of the project, the company began its own farming operation. The ranch hands had no farming skills, so Bosler brought a farm crew from Carlisle, Pennsylvania, to help with leveling of land and planting, and he also hired irrigators from the Greeley colony in Colorado. Although the farming operation was a good marketing tool, it lost money and was a further burden to the company.

Initial land sales were good, and in the first two and a half months, 4,600 acres sold; by the end of 1911, nearly 6,000 acres were in the hands of settlers, representing a gross sales value of over $250,000. From a cash flow perspective, the bad news was that nearly $160,000 of this total had not been collected, and there was not enough cash to repay the short-term loans Bosler had used to finance the summer's expenditures. These notes began to fall due in August, in increments of $25,000.[25]

In order to pay off the early loan installments, Bosler used the proceeds of the year's cattle sales on the ranch and raised another

---

[24] The Bilby tract was colonized by the Cedar Rapids Land Company. *Cheyenne Daily Leader*, 5 Oct. 1906.

[25] Frank C. Bosler to Horace Plunkett, 20 Sept. 1911, Bosler collection.

$125,000 by selling a 25 percent interest in the Rock Creek Company to his cousin Abram. As usual, he did this without consulting with Plunkett, which was bad enough, but he still needed more money, and later in the year he asked Plunkett to advance an additional $50,000. This time, Bosler had gone too far, for Plunkett he had reached his own credit limit with the Omaha bank.[26]

Frustrated at his inability to escape from Bosler's demands, Plunkett angrily upbraided Bosler for making decisions without consulting him. Plunkett was particularly angered by Frank's deal with his cousin. He said, quite incorrectly, that the arrangement with Abram effectively transferred $19,000 from his pocket, adding that under British law the transaction would be as illegal as it was immoral. As always, this elicited a lengthy response from Bosler, who declared that Plunkett's complaints "quite took the starch out of me." Frank Bosler wrote pages to placate Plunkett, but he was not repentant, for when costs again exceeded expectations, he went back to Abram for more money, on the same terms as before.[27]

Of course, Bosler was the majority shareholder in the Diamond Company and could outvote Plunkett on any issue, but he wanted to placate the Irishman to keep his money in the enterprise, which was larger than the Boslers could afford to finance alone. So it was that in the following year, Frank Bosler told Plunkett of another $20,000 obligation, and Plunkett loaned the company another $12,000, saying in a telegram, "relying on you for fair adjustment."[28]

In 1912, Bosler continued the rapid pace of spending on irrigation projects. The Peirce reservoir was completed, and the Bosler and Sand Lake reservoirs were surveyed. Bosler added to the list of capital requirements his new plan for the construction of a hydroelectric plant at the Rock Creek reservoir site. Finally, the company again exhausted its cash resources, and construction of the

---

[26] Plunkett diaries, 29 Oct. 1909, 20 Jan. 1911, and 24 Feb. 1911; Frank C. Bosler to Horace Plunkett, 20 Sept. 1911, and Horace Plunkett to Frank C. Bosler, 18 Oct. 1911, Bosler collection.

[27] Plunkett's interest in the company was not diluted by Abram Bosler's investment, as this investment increased the value of the company by at least as much as Abram Bosler's stock interest. Horace Plunkett to Frank C. Bosler, 4 Jan. 1912, Bosler collection.

[28] Horace Plunkett to Frank C. Bosler, 1, 4, 8, and 9 Jan. 1912 (written from Battle Creek, Mich.); and Frank C. Bosler to Horace Plunkett, 6 Jan. 1912, Bosler collection.

Bosler reservoir had to be suspended for a time. Bosler estimated that all of the new construction would entail $500,000.[29]

This total was large enough, but in his headlong rush to spend money before it could be raised from permanent sources, Bosler burdened the parent company with $500,000 in notes outstanding, mostly short term. He now promised to buy additional stock by mortgaging one of his ranches, but even this source of money was inadequate. This time, he proposed to give Plunkett a "bonus" of stock in the Rock Creek Company for additional advances to the parent company. Plunkett agreed to loan the company an additional $34,000, selling some stock he held in New York to raise the money. Once again, he hoped the project could now be completed, eliminating the cause of all their arguments. In a letter to Johnny Peirce, Plunkett declared, "We shall now have the best lay out in the West for settlers who have the sense to go where they can get all the water they want."[30]

This optimism was soon shattered, for none of Bosler's revenue projections was sound, expenses were higher than he had expected, and the agreement with Plunkett definitely did not put an end to pleas for money. Farming operations lost money, land sales were disappointing, and predictably in 1913 Bosler asked Plunkett for an immediate advance of $16,000 and a further $1,200 at the end of each month thereafter.[31]

This request touched off another sharp response from Plunkett, who now recommended that the company sell its herd of some 7,000 head and concentrate on making land sales. This recommendation was painful for Plunkett, because it eliminated two of his reasons for investing in the Diamond Ranch. Without the cattle, the romance of the range would be gone; in addition, Johnny Peirce would no longer be there to keep an eye on the Boslers.[32]

---

[29] Frank C. Bosler to Horace Plunkett, 9 and 20 March 1912, Bosler collection.

[30] Plunkett was also to receive 400 shares of the Rock Creek Company (nominally $40,000), and he was considering taking additional stock for his outstanding notes with the Diamond Co. Frank C. Bosler to Horace Plunkett, 9 March 1912; Abram Bosler to Horace Plunkett, 20 July 1912; Horace Plunkett to Conrad Young, 15 March 1912; Horace Plunkett to John Peirce, 23 March 1912; and Horace Plunkett to Frank C. Bosler, 10 March 1912, 15 May 1912, and 26 July 1912, Bosler collection.

[31] Frank C. Bosler to Horace Plunkett, 14 and 31 Dec. 1912; and Horace Plunkett to Frank C. Bosler, 2 Jan. 1913, Bosler collection.

[32] Frank C. Bosler to Horace Plunkett, 13 May 1912, Bosler collection.

Bosler opposed the idea of selling the cattle as the company would then be solely dependent on land sales to meet its obligations. Moreover, without the cattle, there would be no market for the alfalfa raised by the farming operations. He also argued that the cattle operation was more profitable than it appeared on the books. In his reply, Plunkett barely restrained his anger. He acknowledged Bosler's argument that selling the cattle would give up income. Then he added, "Your alternative is that I have to pay $16,000 down and $1,200 a month thereafter, a simple solution but one which I cannot adopt, and which I do not think ought to be urged upon a partner who has protected a $42,000 interest with $80,000." Nevertheless, these exchanges of letters solved nothing, and Plunkett continued to be dragged along by his partner in this bottomless financial pit.[33]

Fortunately, there were other matters in the United States that were not so frustrating to Plunkett. In the spring of 1912, Plunkett contacted two of Bill Nye's children, who were then living in New York, because Nye's works were now difficult to obtain, and he hoped to have them republished. He got a letter of introduction from someone at Battle Creek to Nye's daughter Winifred Haynes, wife of a prominent physician in Manhattan, and went to see her. After talking with Mrs. Haynes, Plunkett also called Nye's son Frank Wilson Nye, an advertising salesman, and offered to try to get Macmillan to republish Nye's works. This effort bore fruit, and Frank Wilson Nye published *Bill Nye: His Own Life Story* in 1926; in the book, he acknowledged Plunkett's conversation with him.[34]

At the time of Plunkett's 1912 visit to New York, the women's suffrage movement was in ferment there, encouraged by demonstrations in England and actions at the state level in the United States. He attended a society affair hosted by the Mount Holyoke Alumnae Association at Carnegie Hall, where Lady Warwick

[33] In a postscript to the 11 Jan. 1913 letter, Plunkett said, "If you think it better not to show this to anyone, either bring it with you when we meet to discuss finances, or send it back to me." Horace Plunkett to Frank C. Bosler, 8 and 11 Jan. 1913.

[34] Plunkett diaries, 7 March 1912; and Nye, *Bill Nye*, p. x. Edgar Wilson "Bill" Nye died in 1896, and his wife died before 1910; by 1912 the children were widely scattered. Bessie Loring married Eugene A. Pharr, a Louisiana sugar planter; Winifred Louise Haynes and Frank Wilson Nye both lived in New York City; and Douglas Nye died in Rochester, Minn.

spoke for suffrage for women and against child labor and war. The Countess dropped the names of many famous people but gave no startling anecdotes, and Plunkett was bored "to distraction." At the end of the year he again met Jane Addams at Hull House in Chicago, where they also spoke about women's suffrage, with Jane Addams arguing for militancy and Plunkett counseling moderation.[35]

Plunkett's friendship with Theodore Roosevelt continued to bring invitations to visit the former president, and Plunkett went out to Oyster Bay on Long Island, New York, for lunch. Later at dinner, Ida Tarbell joined them. Roosevelt was then in contention with Taft for the Republican nomination, but he told Plunkett he had little hope that he would succeed. Plunkett's objective in this meeting was to get Roosevelt to revive his interest in conservation and country life and to point out publicly that Taft had dropped these issues. Plunkett told Roosevelt that people wanted to know more about the differences between him and Taft, and Roosevelt agreed to have Pinchot and Plunkett prepare a speech, which he would deliver at an early date.[36]

As devil's advocate, Plunkett asked why Roosevelt didn't just say he had changed his mind about running for a third term. Roosevelt replied, "Because I haven't. In 1904 I said I would not be a candidate and condemned third terms. I did not say in 1905 a 'consecutive third term' because that would be tantamount to announcing that I would be candidate in 1912."

Plunkett replied, "Why don't you give this explanation?"

"I have," Roosevelt said.

"Then give it again as clearly as you have given it to me," Plunkett rejoined.

---

[35] Lady Warwick spoke on 12 March. Women were given the vote in school elections in Kentucky, and the Ohio constitutional convention adopted an amendment permitting women's suffrage. A speaker from England likened the militancy of British suffragettes to the Boston Tea Party. The day after the meeting Plunkett attended in New York, women appeared in Washington, D.C., for a hearing on women's suffrage. *New York Times,* 8, 10, and 12–14 March 1912; and 23 Dec. 1912.

[36] Ida Minerva Tarbell, a muckraking journalist and lecturer, wrote a famous two-volume history of the Standard Oil Company that contributed to public outcry leading to the breakup of the company in 1911. She was born in 1857, and at the time of her meeting with Roosevelt and Plunkett she had just completed her latest book, *The Business of Being a Woman,* in which she argued that women's best contribution was with home and family.

Plunkett said Roosevelt was "greatly chastened" and knew that he had made a huge blunder, but he said he had no choice as he was drawn into it. Roosevelt let Plunkett do most of the talking, which was a great change in their relationship, and Plunkett thought this might mean Roosevelt would do "a little more thinking." "He is a shrewd demagogue with honest purposes & great courage. His ends justify his means . . . & he sometimes is quite unfair to his friends. He may be President again as the masses expect him to make 'howlers' & don't look closely into his 'rekkord.'" The American electorate did not confirm Plunkett's sentiments, for although Roosevelt outpolled Taft in the 1912 election, the split of the Republican Party handed Wilson the presidency with only 42 percent of the popular vote.[37]

Back in America at the end of 1912, Plunkett made his usual trips to Omaha and Cheyenne to look after his business interests. He found that his team in Omaha was functioning well, but the earnings of £5,000 were swallowed up by the cash needs of the Diamond Ranch and Wheatland Project in Wyoming. Plunkett extended the profit-sharing arrangement in Omaha he had made with Young to the understudy, Selwyn Doherty. Also in Omaha, Plunkett attended a farmers' convention and spoke, happily noting in his diary that "Plunkettism" was growing in that area, with better results from much less effort on his part than was the case in Ireland.[38]

Plunkett ended the year 1912 as he had begun it, at the Battle Creek Sanitarium, where he had four days of X-rays. From Battle Creek, he went to Madison, Wisconsin, to speak to the state legislature about rural life. From there, he went to Chicago, where he met with Cyrus McCormick to induce the International Harvester Company to support his work. In late January, the Southern Commercial Congress promoted the idea of a delegation from the United States to Europe to look into the matter of rural credit and asked Plunkett to address their meeting in Washington. Plunkett persuaded them to broaden the scope of the inquiry to include

---

[37] Plunkett diaries, 14 May 1912. Wilson received 435 electoral votes, Roosevelt received 88 from six states (California gave two of its thirteen votes to Wilson), and Taft received 8.

[38] Doherty was to receive 2 1/2 percent of profits, and in 1912 he was given a $100 bonus in addition to his $500 salary. Plunkett diaries, 14, 16–17, and 23 Dec. 1912.

business organization of farmers. President Taft addressed the meeting and used the occasion to publicly acknowledge Plunkett's services, needling Roosevelt at the same time.[39]

In New York before sailing for home, Plunkett lunched with Theodore Roosevelt's cousin, Franklin Delano Roosevelt, who had been appointed assistant secretary of the navy by President Wilson. Plunkett thought FDR was a politician "of a rather uppish conceited sort," whose redeeming feature was that he had a "nice wife." Later, he spent a couple of hours at Oyster Bay with Theodore Roosevelt.[40]

Plunkett was now a confirmed disciple of the Battle Creek Sanitarium method, and back home in Ireland in the spring of 1913, he delivered a lecture to the Royal Dublin Society entitled "Some American Thoughts upon Health," describing his experiences at Battle Creek. He praised the diagnostic procedure and the use of exercise and diet (chiefly vegetarian) to treat physical disorders. A month after this lecture, he hosted Dr. William H. Riley and his wife, Dr. Mary Riley, from Battle Creek and invited a local doctor and a professor to meet with them. The resulting "symposium" was interesting, and Plunkett thought that the "fashionable" practice of medicine he was accustomed to in Britain did not compare favorably with the Battle Creek system.[41]

While the Rileys were visiting him, Plunkett had an attack of dizziness, and Dr. William Riley gave him a thorough examination. Dr. Riley said that Plunkett was in imminent danger of serious complications that could be fatal, but despite this dire prediction, the doctor's treatment regimen consisted of only two straightforward instructions. He said Plunkett's body was starved, so that he

[39] Digby, *Horace Plunkett*, 137; and Plunkett diaries, 25 Dec. 1912–19 Jan. 1913. President Taft said in part, "In our workaday politics one gets just a little bit tired of the use of the term Progressive by gentlemen who work no progress except for platform purposes, and so when we meet a man who has made progress for the people such as we are all seeking, he is entitled to our respect. . . . We have a great deal to learn, and I have no doubt that from such men as Sir Horace we can learn a great deal. He has shown by what he has done, and not by what he has said, that he is a real progressive." *New York Times*, 28 Jan. 1913.

[40] Plunkett diaries, 2 Feb. 1913. Franklin Delano Roosevelt was assistant secretary of the navy from 1913 to 1920.

[41] Plunkett diaries, 29–30 April 1913. The lecture was delivered 12 March 1913 before the Royal Dublin Society. Plunkett, *Some Tendencies of Modern Medicine*, 11.

must eat more but keep his diet as near to "non-toxic" as possible, eating only so much chicken, bacon, and ham as necessary to stimulate appetite. He must be out of doors as much as possible and above all "*rest, rest, rest.*"[42]

Undoubtedly encouraged by the Drs. Riley, Plunkett arranged to have his Royal Dublin Society lecture published in the *Lancet* in London and also separately as a pamphlet. For the U.S. audience, he reprinted the lecture as a pamphlet with introductions by a prominent New York physician and a professor at Yale. In August, Plunkett hosted Dr. Charles E. Stewart and his wife from Battle Creek, but it is unknown if this was a social call or another transatlantic house call.[43]

---

[42] Dr. Riley did not promise that his treatment would relieve Plunkett's persistent tinnitus, which he had no opinion on. Plunkett diaries, 1 May 1913.

[43] The *Lancet* article was dated 24 May 1913, and the London pamphlet was published by Eason and Son. Plunkett's introduction to the American pamphlet was dated 1 Aug. 1913 and was published by Good Health Publishing Company of Battle Creek, a company headed by Dr. John Harvey Kellogg, with his brother William Keith Kellogg as business manager. Also Plunkett diaries, 19 Aug. 1913.

12

Advisor to Another President

WHILE HE WAS STILL in the United States in early 1913, Plunkett went to Princeton, New Jersey, to try to open communications with Woodrow Wilson, the former New Jersey governor who had just been elected president. Wilson listened attentively, seemed interested, and told Plunkett to write "anything that occurred to me from time to time." Later, Wilson sent Plunkett a letter endorsing the idea of establishing an American Commission on Agricultural Credit.[1]

In April, David Franklin Houston, Wilson's new secretary of agriculture, named Dr. Thomas N. Carver, a Harvard professor of economics, to direct a rural organization service, and this move impelled Plunkett to take up President Wilson's invitation to give him advice. Fearing that Wilson's Department of Agriculture would not carry out his concept of rural cooperative work, he cautioned the president against letting the department organize the cooperative movement in the United States.[2]

Plunkett's efforts to influence the Wilson administration were considerably enhanced by his introduction to Col. Edward M. House, who had Wilson's ear even though he did not have a formal position in the government. Walter Hines Page, who arrived in

---

[1] Plunkett diaries, 21 and 31 Jan. 1913. The commission, which had the fulsome title of Permanent Commission on Agricultural Organization, Cooperation and Rural Credits, sailed for Europe in late April. *New York Times,* 26 April 1913.

[2] Plunkett diaries, 17 and 29 May 1913; and *New York Times,* 8 April 1913. David Franklin Houston, born in 1866, was chancellor of Washington University in St. Louis when he was appointed secretary of agriculture to replace James Wilson, the secretary under McKinley, Roosevelt, and Taft.

London as the new American ambassador in the summer of 1913, suggested that Plunkett should see House and apparently introduced the two men. In their first meeting in London in June, Plunkett and House obviously talked about Plunkett's letter to President Wilson regarding the Department of Agriculture. House urged Plunkett to go to America soon to give his ideas on rural organization to his friend Secretary Houston, and Plunkett immediately made plans to do so.[3]

In July, the Permanent American Commission on Agricultural Organization, Cooperation and Rural Credits arrived in the United Kingdom as a part of its tour to investigate the adaptability of European systems to American agriculture. The commission had 100 members, including a number of Plunkett's American friends, and when they came to Dublin, the viceroy and Lady Aberdeen greeted them "in full state." Afterward, the commission retired to Plunkett House, where Horace Plunkett greeted them and delivered a speech outlining his cooperation ideas. Among Plunkett's guests on this occasion were old Henry Wallace of *Wallace's Farmer* and James Wilson, the former secretary of agriculture. With Wallace was his son, Henry Agard Wallace, who would later become secretary of agriculture during Franklin Delano Roosevelt's first term and vice president in 1940.[4]

After the American commission left Ireland, Dr. Thomas Carver, Charles McCarthy, and three other rural economists came to visit Plunkett, apparently at the behest of Colonel House. Plunkett spent a day arguing with Carver, who had written a book on rural economics that did not even mention the concept of coopera-

---

[3] Walter Hines Page, who was a partner of Doubleday, Page & Co. and editor of *World's Work* magazine when he was appointed ambassador, was presented to the King on May 30. Colonel House played a central role in the choice of Wilson's appointees, including both Houston and Page, and it would have been natural for Page to introduce Plunkett to House. Plunkett was in England at the time and met Page on June 2. After seeing House, Plunkett wrote Bosler that Wilson's agricultural policy was "very good and altogether along my lines." Horace Plunkett to Frank Bosler, 25 June 1913, Bosler Collection; Hodgson, *Woodrow Wilson's Right Hand*, 79–80; and Plunkett diaries, 2 and 19 June 1913. Dr. Carver's visit to Dublin is mentioned in Horace Plunkett to Frank Bosler, 23 Aug. 1913, Bosler collection. The only public office House occupied was as Wilson's special representative in Paris and as a member of the American delegation to the Paris peace conference.

[4] Sen. Duncan Upshaw Fletcher of Florida chaired the commission. *New York Times,* 26 April 1913 and 8 and 13 July 1913; and Plunkett diaries, 7–24 July 1913.

tive organizations. "Ergo," Plunkett groused in his diary, "he must oppose the Irish idea." Plunkett called Carver "stupid" and said he was "the most unfortunate selection for the headship of the rural organization service." Nevertheless, the other men with Carver seemed inclined to Plunkett's point of view, and Plunkett was optimistic that Carver would change his mind later.[5]

Plunkett landed in New York in October 1913 and had dinner with House before going to Washington. This time, House did most of the talking, telling Plunkett how he had controlled the Texas legislature for years, and Plunkett gained the impression that House "knew a great deal about Mexico." We do not know how much House divulged in their conversation, but Plunkett's summary considerably understated the facts. House was a part of the group of influential Texans who were deeply involved in American financial expansion in Mexico, and his brother was a director of seven American mining companies in Mexico. These American interests argued for support of Carranza (who favored protection of private property) and for intervention against the Huerta government.[6]

Plunkett checked House's credentials with Henry Smith Pritchett, president of the Carnegie Foundation, but the only detail from Pritchett that Plunkett recorded was that House had made a good deal of money in Texas and "never bribed anyone." Later, in Washington, Plunkett discussed agricultural organization and thought he had cleared up much misunderstanding. In December, he wrote to House, praising the work of Secretary Houston and emphasizing the need to devote attention to the matter of rural credit.[7]

The U.S. government recognized and appreciated Plunkett as an expert on agricultural problems, but his relationship with Colonel

---

[5] With Carver and McCarthy were Bradford Knapp, chief of farm demonstration work in the U.S. Bureau of Plant Industry, and two unnamed men. Plunkett diaries, 21 and 22 Aug. 1913. Carver's book, which so offended Plunkett, was *Principles of Rural Economics*.

[6] Plunkett diaries, 12 Oct. 1913. When the Hearst newspapers circulated stories about House's personal interests in Mexico, House claimed he did not have "a dollar" invested there. Nevertheless, he was a shareholder of the Texas Company and had other close connections with oil interests in Mexico. Hodgson, *Woodrow Wilson's Right Hand*, 87; and Hart, *Empire and Revolution*, 162, 306.

[7] Plunkett said House was quiet and shrewd, with a great reserve of strength and an impressive personality, although he was "undersized" and had a weak chin. Plunkett diaries, 17 Oct. 1913. Plunkett's letter of 16 Dec. 1913 to House is quoted in Digby, *Horace Plunkett*, 138–40.

House also gave him the opportunity to advise on matters involving foreign affairs. This opportunity arose in connection with Wilson's controversy with Mexico, which also involved U.S. relations with the United Kingdom. Gen. Victoriano Huerta assumed the presidency of Mexico in February 1913 at the end of Taft's term in office, only days before Wilson was sworn in. The U.S. ambassador to Mexico applauded Huerta's action and urged the diplomatic corps in Mexico City to recognize the new regime. Although his own government had not yet acted, the ambassador's action was consistent with the settled policy followed heretofore by the United States in such instances of a regime change.[8]

Shortly after his inauguration, Wilson asked the British government what it intended to do about recognizing Huerta and received the response that they would not recognize him. However, some three weeks later, the British did recognize the new government "provisionally," saying they had changed their mind. This abrupt change, which was made without consulting the United States, made Wilson suspicious that the decision had to do with large British investments in Mexico. Wilson, hoping to force Huerta to resign, then declared that he would not recognize "a government of butchers." Instead of resigning, Huerta arrested all the members of the Mexican chamber of deputies and became military dictator of the country.[9]

The friction between the United Kingdom and the United States caused by the Huerta coup came on top of another issue between the two countries. This problem dated back to a 1912 U.S. law granting American coastwise vessels exemption from tolls in the new Panama Canal (which was scheduled to open in 1914). The British protested that the exemption was contrary to the second Hay–Paunceforte Treaty, which guaranteed all nations equal access to the canal during peacetime.[10]

---

[8] Hodgson, *Woodrow Wilson's Right Hand,* 87. Henry Lane Wilson was appointed as ambassador to Mexico on Dec. 21, 1909, and diplomatic relations were interrupted on Feb. 18, 1913. Wilson left Mexico on July 17, 1913, and diplomatic relations were not resumed until 1917.

[9] Knock, *To End All Wars*, 25. Weetman Pearson, later Lord Cowdray, had large investments in railroads and oil properties in Mexico. Wilson's query regarding British policy on recognition of Huerta was handled in London by Irwin Laughlin, charge d'affaires, because Walter Hines Page, the newly appointed ambassador, had not yet reached London.

[10] The second Hay–Paunceforte treaty between the United States and Great Britain conceded the right of the United States to build and fortify the canal. It was ratified on 21 Feb. 1902.

From New York, Plunkett went to Washington, D.C., where he saw his friend young Tom Spring-Rice, who was a cousin of the British ambassador to the United States. Plunkett outlined for Spring-Rice his own plan for easing difficulties between the two countries. He suggested that the United Kingdom support the United States in Mexico as an inducement for the Americans to yield to the British position on the canal tolls. The next day, he discussed the Mexican situation with House and undoubtedly repeated the suggestion he had given Spring-Rice. Considering House's personal interests in Mexico and the fact he was involved in making Wilson's policy on the matter, it is not surprising that he was unimpressed with Plunkett's idea. Indeed, he told Plunkett that U.S. intervention in Mexican affairs would require less time, money, and men than most people thought.[11]

Four days after his interview with House, Plunkett was able to see the president, who was looking "wretchedly ill," undoubtedly because of the mess with Mexico. We do not know if Plunkett repeated to President Wilson what he had told young Spring-Rice.[12]

In the end, the British government did withdraw its pro-Huerta minister from Mexico and support Wilson's efforts to remove Huerta. For his part, Wilson abandoned his campaign pledge to support American exemption on canal tolls. It is easy to suggest that Plunkett whispering in the ears of the British diplomats had brought about this happy change of affairs, but that was hardly the case. In fact, the two countries were then in what a historian has called the "great rapprochement," which became more apparent when German aggression brought on the First World War.[13]

The British foreign secretary sent his private secretary to Washington, D.C., in late October, ostensibly on a "private visit," which

---

[11] Plunkett diaries, 15 Oct. 1913; and Digby, *Horace Plunkett,* 138.

[12] Although Plunkett did not know it when he made his suggestion to Spring-Rice and Wilson, the groundwork for the type of settlement he proposed had already been laid by House when he met Sir Edward Grey in London on 3 July. At that meeting, House confided to Grey that the president would move to have the American preference on canal tolls repealed. Nevertheless, the matter continued to irk the British until Congress acted, and Wilson did not request the repeal until the beginning of March 1914. Hendrick, *Life and Letters of Walter H. Page,* vol. 1, 245–46, 253.

[13] The "great rapprochement" is the term suggested by Bradford Perkins and quoted in Ferrill, *American Diplomacy,* 492–94. After months of rumors beginning in January 1914, Sir Lionel Carden, the pro-Huerta British minister to Mexico, was finally officially recalled on 15 April 1914. *New York Times,* 6 Jan. 1914 and 16 April 1914.

nonetheless involved at least two meetings with Wilson. It was obvious that the two countries were looking for common ground in their diplomatic difficulties, and in particular, Wilson agreed that the treaties regarding canal tolls must be honored. Under the circumstances, Plunkett's voice may not have mattered much. In any case, when Plunkett saw House again in mid-December, the latter told him that relations between the United Kingdom and the United States were now "perfect."[14]

While in Washington, Plunkett had time for some private diplomacy, as well. He dined with Florida senator Fletcher and his wife, whose daughter was engaged to Plunkett's private secretary in Ireland, Lionel Smith-Gordon. Smith-Gordon's mother was opposed to the match, and Plunkett took the opportunity to have a "heart to heart" talk about the relationship with Mrs. Fletcher.[15]

From Washington, Plunkett went back to Battle Creek, this time for 44 days. He had more X-rays of the heart and lungs and had some blood tests and surgery. In early December, he set out for Cheyenne, and on the way a high plains winter snowstorm halted the train at Sidney, Nebraska, forcing him to spend the night on the train. The next day at Cheyenne he found twenty inches of snow on the ground. Robert Carey was snowed in at his ranch and could not meet with Plunkett, destroying a chief reason for the trip west.[16]

If possible, Plunkett's relationship with Bosler and the Diamond Company was even more frustrating in the next two years than it had been in the past. In the fall of 1913, Plunkett was alarmed to

---

[14] Sir Cecil Spring-Rice, the British ambassador to the United States, had been ill during the summer of 1913 and living in Dublin, N.H. It was therefore easy for the British foreign secretary to send Sir William Tyrrell to the United States on the pretense of visiting his friend Ambassador Spring-Rice without raising any questions about the reason for the visit. Later, Tyrrell admitted he had spoken "frankly" about Mexico with Wilson. *New York Times,* 23 Nov. 1913. Also, Plunkett diaries, 17 Dec. 1913; and Hodgson, *Woodrow Wilson's Right Hand,* 89.

[15] Lionel Eldred Pottinger Smith-Gordon was the son of Sir Lionel Smith-Gordon, who was a member (representing Canada) of the commission that visited Ireland earlier in the year. Lady Smith-Gordon was distressed at the engagement of Nell Fletcher to young Lionel because she feared the girl would not have children and the family name would then die out. Plunkett diaries, 28 Sept. 1913 and 15 and 17 Oct. 1913.

[16] Plunkett's surgery was a painful procedure to relieve an enlarged prostate and to remove three "large" hemorrhoids and some "superfluous" tissue. Roosevelt's friend Jacob Riis and two prominent Michigan grangers came to visit Plunkett at Battle Creek. Plunkett diaries, 18 Oct. and 5–7 Dec. 1913.

learn that the new settlers who had purchased land were having difficulty meeting their payments to the company. The sales manager recommended that the company hire the settlers for construction work, and Plunkett endorsed the recommendation. Bosler met the company's 1913 cash emergency by borrowing $60,000 from the livestock commission house against cattle shipments for the year. Once again, he promised to sell one of his ranches to raise more money for the company.[17]

For the year 1913, cattle sales were $85,000, and the cattle operation actually showed a profit of $35,000, permitting a $17,000 reduction in the company's floating debt and a $5,000 payment on the mortgage acquired with the Taylor ranch. Bosler was confident that a "fair" amount of the arrears on sales contracts was collectible. All of this sounded fairly optimistic, until Plunkett's auditor, T. C. Cannon, pointed out that Bosler had taken a large amount of land in trade on sales to settlers. These properties were scattered through five states, and some of them were encumbered, making it even more difficult for the company to realize on its claims.[18]

Bosler now used a new tactic to extract money from Plunkett. Although he seemingly had the money to continue with construction, he told Plunkett that he could not go forward because the company had a $100,000 loan that might be called for payment at any time. Therefore, he proposed to defer construction and apply all available funds to reduce debts if Plunkett refused to advance any more money. This action would limit the amount of irrigation water and stop land sales. This dire prediction was a prelude to Bosler's next scheme, which was to have Plunkett convert his notes to preferred stock, which would not have to be repaid on a fixed schedule.[19]

---

[17] By the middle of Oct., the debt to the Smith Brothers Commission Co. had been reduced to less than $13,000, but Bosler did not expect to have enough additional sales to pay it off. By the fall of 1914, Bosler was still hoping to sell the Nance County lands. Horace Plunkett to Frank C. Bosler, 3 March 1913; and Frank C. Bosler to Horace Plunkett, 22 April 1913, 13 June 1913, 13 Oct. 1913, and 17 Aug. 1914, Bosler collection.

[18] The land taken in trade was in Kans., S. Dak., Neb., Mo., and Iowa; there were two lots in Neb. and a factory in Iowa. Horace Plunkett to Frank C. Bosler, 17 Dec. 1913, Bosler collection.

[19] The $100,000 obligation was owed to Frank O. Harrison and the Cooper heirs, on the Cooper ranch purchase. Bosler offered to convert his own notes to stock, as well. Frank C. Bosler to Horace Plunkett, 31 Dec. 1913 and 20 Jan. 1914; and F. O. Harrison to Horace Plunkett, 1 Feb. 1914, Bosler collection.

Plunkett, now used to Bosler's style, calmly rejected the preferred stock proposal but agreed to take a four-year note for $60,000, provided interest was paid currently, and he arranged to extend or pay the company's $20,000 note to the Nebraska National Bank. Once again, he tried to control Bosler's financial excesses by having him agree to specify future financing in detail.[20]

A month later, Bosler was back with a proposal to have the company issue $200,000 of fifteen-year bonds. Even though he had not been able to sell bonds on his Wyoming ranching properties, Bosler thought this bond issue would be "very conservative" and presumably saleable. Plunkett replied that this was a "splendid" plan, and Bosler at once began arrangements to issue the bonds. Unfortunately, when this proposal was placed before the real world of finance in the United States, Bosler could not get a rate below 8 percent, and the matter then languished for a time.[21]

Having proposed a way to reduce the Diamond Company's debts—although without actually reducing them—Bosler now embarked on a plan to spend more money. He proposed to buy 2,000 head of cattle for about $100,000, in the expectation that the company could make a profit of $20,000 on the investment. Having failed to sell the bonds, he set out to borrow $150,000 at 8 percent to buy cattle, which he optimistically assumed would return 30 to 40 percent of their purchase price the first year. Of course, Plunkett had the financial scars from investing for quick profits in cattle, and he was cool toward this new plunge into the cattle business. He said, "My mind has not recovered from the shock caused by the slump in [cattle] prices thirty years ago."[22]

The succession of Bosler financing schemes continued in the middle of 1914, this time with a proposal for a $300,000 5½ percent mortgage loan. Bosler asked Plunkett to try to sell this idea to the "proper" people in the United Kingdom, but nothing came of it. In the fall of 1914, Bosler asked Plunkett to defer receipt of his interest

[20] Plunkett also asked Bosler to guarantee that the security for his new note not be diminished while it was outstanding. Horace Plunkett to Frank C. Bosler, 15 Jan. 1914, Bosler collection.

[21] The issue Bosler prepared was for $200,000 of twenty-year (not fifteen-year), 6 percent bonds, with sinking fund payments of only $3,000 per year. Frank C. Bosler to Horace Plunkett, 16 Feb. 1914 and 13 March 1914; and Horace Plunkett to Frank C. Bosler, 4 March 1914, Bosler collection.

[22] Frank C. Bosler to Horace Plunkett, 21 April 1914; and Horace Plunkett to Frank C. Bosler, 21 May 1914, Bosler collection.

payments for the year so that wages could be paid and supplies purchased, and Plunkett agreed to take a one-year note for this $5,900 item.[23]

In order to sell bonds secured by a mortgage on the company properties, Bosler had to remove a $100,000 mortgage on one of the ranches. Bosler asked the mortgagees, who were English, to accept the new company bonds in payment of their mortgage, but they rejected this offer and demanded cash. Once again, Bosler had to pay off a long-term obligation with short-term borrowings. Moreover, in spite of all these efforts, the year 1915 passed without the sale of any bonds.[24]

In the fall of 1915, Plunkett told Bosler that a change in the British tax law made it impossible for him to continue to advance money on the Diamond Company projects. This time, Plunkett's request resulted in serious negotiations to extricate himself from the Diamond Company morass. These negotiations revealed that the Diamond Company had undistributed earnings of some $266,000, and Plunkett claimed his share of this amount, which together with the original value of his stock and the principal and interest on his notes totaled $145,000.[25]

Although Plunkett frequently complained about Bosler, it is amazing that he still let the latter entice him to invest in mining ventures. The Boslers were large shareholders in the Carter Mining Company, which held mining claims near Ohio City, Colorado, and Plunkett bought 4,000 shares that cost him $3,000. The company spent some $900,000 by the beginning of 1915, and Plunkett finally characterized the operation as "looking for metal & finding stone," which is perhaps the best summary of it.[26]

---

[23] The prospectus for the bond issue showed that ranch operations made nearly $55,000 in 1913, and the Rock Creek Company earned an additional $18,000, for a total of $73,000. Frank C. Bosler to Horace Plunkett, 7 and 14 July 1914; and Horace Plunkett to Frank C. Bosler, 27 July 1914 and 5 Aug. 1914, Bosler collection. Also, Plunkett diaries, 21 Dec. 1915.

[24] Horace Plunkett to Frank C. Bosler, 16 Sept. 1915; and Frank C. Bosler to Horace Plunkett, 5 Dec. 1915, Bosler collection.

[25] Horace Plunkett to Frank C. Bosler, 5 Aug. 1915 and 22 Dec. 1915, Bosler collection.

[26] *The Carter Mining Company President's Report,* 2 Jan. 1915; Frank C. Bosler to Horace Plunkett, 4 Oct. 1913 and 8 Jan. 1915; and Horace Plunkett to Frank C. Bosler, 11 Jan. 1915, Bosler collection. Still another of Plunkett's mining ventures was the Ashland Mining Company, where Bosler was also apparently involved with Plunkett. At the end of 1914, Plunkett complained that he had heard nothing from the company for a long time—always a bad sign in the case of a mining speculation. Horace Plunkett to Frank C. Bosler, 21 Dec. 1914, Bosler collection.

Plunkett's other long-term investment, the Wyoming Development Company, finally issued liquidation certificates covering its land sales contracts in the spring of 1914. Plunkett's certificate was for nearly $193,000, but he was pessimistic about the probability of collecting that amount. When he transferred his shares in the two Wheatland companies to his Nebraska holding company, he wrote down their value, but it is unclear how much he considered the loss to be.[27]

The assassination of Archduke Francis Ferdinand at the end of June 1914 claimed everyone's attention as war broke out. Plunkett did not immediately make an effort to influence American government policy toward the resulting war, and like so many Europeans on both sides of the conflict, he expected that the war would end "next summer." With that prospect in view, the Allies did not want the Americans interfering with their plans to impose peace on the enemy. Before long, though, the reality of the situation began to sink in with Plunkett, and by late October he wrote, "All my talks with people who think, persuades me that nothing will be tomorrow as it is today. The world is in dissolution." It was becoming increasingly obvious that the need for American help outweighed the disadvantages of American meddling in European affairs, and Plunkett began to think of ways to influence American policy toward the war.[28]

Soon he was insinuating himself into the inner counsels of the British and American governments, as they dealt with their evolving policies in the world's first general war. It was a heady experience for Plunkett, who was thrilled when he was invited to hear secret high-level conversations between the two governments and to give his opinions on their deliberations.[29]

---

[27] In a letter to Conrad Young in the fall of 1917, Plunkett said he expected a "heavy" loss, which he attributed to "slackness in administration." Horace Plunkett to Conrad Young, 30 Oct. 1917, Plunkett Collection, Nebraska State Historical Society (hereafter cited as NSHS collection).

[28] Horace Plunkett to Frank C. Bosler, 23 Oct. 1914, Bosler collection; and Plunkett diaries, 3 Sept. 1914 and 24 Oct. 1914. Among the Allies, the British position was further complicated by the British self-governing colonies, who had their own designs on the overseas territories of the Central Powers (e.g., Australia did not want to give up New Guinea), and the colonies' demands had a powerful influence over British government policy. Although the Allies came round to the need for American help, there was still reluctance to accept actual American intervention in the war.

[29] Plunkett diaries, 4 June 1915.

Reaching U.S. policymakers was not an easy task, and Plunkett never achieved the intimacy with President Wilson that he enjoyed with Theodore Roosevelt. Indeed, despite Wilson's cordiality and the occasional warm letter of thanks, Plunkett seldom saw the president face to face. There were few White House meetings and no intimate dinners with the president's family, so that Plunkett's ideas—and he had many during the Wilson presidency—reached Wilson's ears only indirectly, usually through Colonel House.

Before making his annual journey to the United States at the end of 1914, Plunkett told the British foreign secretary, Sir Edward Grey, that he knew half of Wilson's cabinet as well as other U.S. leaders. There followed a remarkable conversation in which Plunkett, playing devil's advocate, criticized Grey's handling of the events leading up to the war and asked Grey to defend his actions. Although Grey seemed to respond in a fulsome manner, he did not confide any secret information to Plunkett, and at times he noted that there were other factors and other information that he could not disclose. The interview apparently convinced Plunkett that Grey had taken him into his confidence, but in fact, Grey never confided official secrets to the Irishman, although other British officials sometimes did.[30]

In the early days of the war, both the Allies and Washington, D.C., spent a good deal of time considering what sort of peace terms to impose on the Central Powers—who, of course, were far from defeated at that point. In Washington, Plunkett tried to learn the American position on peace talks. Sen. Elihu Root was strongly opposed to "premature" peace talks, but House told Plunkett that the president did have peace proposals in mind, which had apparently been floated by Germany. Plunkett argued strongly with House against any settlement that did not acknowledge Germany as the aggressor, and he later left a message at the British embassy warning the ambassador of German-inspired peace proposals.[31]

After a quick business trip to the West, Plunkett went to Battle Creek, and as usual, Dr. Kellogg took advantage of the presence of

[30] In typical fashion, Plunkett noted in his diary that Grey was not a "big" man intellectually, that he lacked imagination, "& I should say his psychology was not for diplomacy." Plunkett diaries, 7 Dec. 1914.

[31] Ibid., 21 and 22 Dec. 1914.

his important guest to have him lecture on "War and Civilization." From Battle Creek, Plunkett turned to fund-raising for cooperative organization, this time not for Ireland but for American cooperatives. He approached both Julius Rosenwald of Sears & Roebuck and the Carnegie Foundation, receiving $5,000 from the latter. Then it was back to see House again.[32]

House assured Plunkett that Wilson had to appear to be pro-German in order to demonstrate to the Germans that he was observing strict neutrality, and Plunkett passed this observation on to Grey. Given that House was freely spreading these words of assurance about Wilson, Grey must have received this sort of "inside" information from more than one source. House also told Plunkett "many things" that he considered too confidential to commit to writing—a comment that appeared often in his diary entries involving House interviews.[33]

From New York, Plunkett went to Boston to talk with two Harvard professors. A. Lawrence Lowell was professor of government and a friend from previous visits, but this time Plunkett wanted to talk about the war, not farm policy. Prof. Archibald Cary Coolidge, a cousin of the future president, was a historian who had been an exchange professor in Berlin before the outbreak of the war in Europe. Both men gave Plunkett their assessment of public opinion in America and ideas for the ultimate peace settlement in Europe. Plunkett took notes—in the case of Coolidge, too extensive to record in his diary—and then he was off again to Washington, D.C.[34]

In Washington, Plunkett met Ambassador Sir Cecil Spring-Rice at the British embassy, and in this conversation he tried to moderate the British attitude vis-à-vis the United States: "I pressed our ambassador to put our dignity in our pockets & make

[32] Ibid., 19 Dec. 1914 and 1–11 and 13 Jan. 1915.

[33] While he was in New York, Plunkett also gave a ten-minute speech in support of fund-raising for the Prince of Wales Relief Fund. Plunkett diaries, 11 and 12 Jan. 1915. Ralph Stuart-Wortley was treasurer for the Prince of Wales fund in New York.

[34] Prof. Archibald Cary Coolidge lectured at the University of Berlin from 1913 until February 1914. He was later a member of "The Inquiry," a group organized in the summer of 1917 at House's suggestion to provide information and advice to Wilson for use in the peace conference at the end of the war. This was a source contributing to Wilson's surprising knowledge of the issues involved in the Balkans, Poland, and the Adriatic. Seymour, *Intimate Papers of Colonel House* vol. 3, 171–72.

the largest concessions to even the bad political susceptibilities & sordid machinations by which Wilson is being dragged into the appearance of pro-German leanings. I urged giving American shipping a lot of rope & then pounce on a suspected cargo."[35]

Plunkett also met Emmanuel Havenith, the Belgian minister to the United States, and told him that Coolidge had argued that Belgium should make a convincing denial of the German claim that there was an Anglo–Belgian conspiracy against Germany. He wanted this denial to come from Belgium rather than the United Kingdom to avoid any backlash of American public opinion against the British. The next day Plunkett again saw Havenith, who was working on a pamphlet to accomplish this purpose. Plunkett thought the pamphlet was "sadly done," whereupon he made some suggestions to make the English more intelligible.[36]

House welcomed Plunkett's opinions, but the U.S. State Department did not regard Plunkett as a diplomatic resource. Plunkett saw Secretary of State William Jennings Bryan, where he found "little knowledge & no thought worth recording," noting that Bryan "platitudinised" him, then said their conversations were not for publication. Plunkett's diplomatic efforts also ran into a stone wall at the White House, where he could not get an appointment with the president. "Wilson feared to see me," Plunkett wrote. This is a strange entry, for Colonel House would normally have arranged Plunkett's appointment with the president. We know Plunkett was in contact with House, because the latter gave him a letter to carry over to Ambassador Page in London, but there is no mention why House did not—or could not—make the appointment with Wilson. Following this rebuff, Plunkett sailed for Liverpool on the *Lapland*.[37]

The *Lapland* crossed the Atlantic on a southerly course to avoid German submarines. Back in London, Plunkett was summoned to see Grey and report on his American conversations, including those with opinion leaders outside the U.S. government. House was also expected shortly in London, and Grey authorized Plunkett to tell House that America could only hope to affect the outcome if she

[35] Plunkett diaries, 16–17 Jan. 1915.
[36] Ibid.
[37] Ibid., 18, 20, and 22 Jan. 1915.

"plays the big part," that is, as a participant in a League of Peace, where all members would agree to submit quarrels to arbitration.[38]

Plunkett was still sufficiently optimistic that England could easily defeat the Germans; hence, he urged Grey to ask the neutral countries, including the United States, only for "endorsement" of British proposals for peace, leaving England free to dominate peace negotiations. He suggested that this position could be made palatable to the United States if the United Kingdom foreswore any territorial claims, and Grey told him this was impossible, because the British had already made commitments to the self-governing colonies. Obviously, Plunkett's view of the true balance of power within the empire was badly out of date, for when he pointed out that the colonies could be asked to make the sacrifice to enable England to secure a lasting peace, he apparently did not get a response from Grey.[39]

When House arrived in England, Plunkett again argued strongly against U.S. offers to arbitrate, "now, or ever upon matters to be settled between them." On the other hand, he urged the United States to be ready to support a policy of lasting peace. House politely did not respond; however, the concept that the United States should support British initiatives but not interfere was unlikely to be well-received in Washington. Indeed, the United States carried on a large trade with all of the belligerents, which made it difficult to be openly pro-British.

Britain, with its large navy, was blockading Germany and hoped to starve her as well as deprive her access to munitions from overseas. The United States, as the largest neutral trading country, was enjoying huge increases in exports to the belligerents and claimed the right to trade freely with both sides. Plunkett pointed out to House that in Britain there was considerable irritation at the "spiritless & selfish" neutrality of the United States (here apparently quoting Theodore Roosevelt), and he said that Russia and France were even more disturbed by it. Plunkett thought House agreed with him on this point.[40]

As early as the end of March 1915, Plunkett finally came to the realization that the Germans would not be easily beaten, and he noted that talk of the Germans being "on the run" was "ignorant gossip." Moreover, the Americans did not think it fair for the

[38] Ibid., 6 Feb. 1915. [39] Ibid. [40] Ibid., 7 Feb. 1915.

Allies to ask for help while at the same time excluding the United States from the peace process. When House returned to London after visiting the Germans, he frankly told Plunkett that little progress could be made in the peace process until the Allies accomplished "something" in the battlefield, as the Germans were satisfied that they were holding the Allies all along the line.

Plunkett now realized that diplomatic efforts with the United States should no longer be aimed at *preventing* U.S. interference, but instead should concentrate on *securing* U.S. support of the Allies, although he still hoped that U.S. troops would not be required. He still clung to the hope that the mere threat of American intervention would bring the Germans to the negotiating table, giving the British a free hand to dictate the terms of peace.[41]

The torpedoing of the *Lusitania* on May 7, 1915, brought much-needed encouragement to the British, who now supposed that the United States would have to enter the war and bring it quickly to an end. Plunkett suggested to House that the United States should threaten the use of an expeditionary force but not actually deploy it, and he said that House agreed that such a threat would "practically settle the business." Later, Plunkett went with House to discuss the idea with Arthur James Balfour, who had just been named first lord of the admiralty.[42]

As House prepared to go back to the United States to consult with Wilson, Plunkett agreed to understudy for him in London and to interview ministers and others there when asked to do so by cable. To facilitate these communications, Plunkett devised a simple substitution code to be used in their cable transmissions, but then he had to get permission to use the code, because only diplomatic officials could send coded messages. To get around this restriction, Plunkett asked for Balfour's help, and the latter's private secretary made the necessary arrangements. Horace Plunkett was now functioning as a peculiar sort of private diplomat.[43]

---

[41] Ibid., 28 March 1915.

[42] Ibid., 7, 8, and 10 May 1915 and 2 and 4 June 1915. Plunkett called the meeting with Balfour "thrillingly interesting." House dramatically read comments from the U.S. ambassador in Berlin, saying that if the British men knew what Germany intended to do with them if she won, they would "come from the graveyards to join the colors."

[43] In the substitution code, "Aaron" was Wilson, "Beverly" was House, "Aftermath" was Balfour, and "Affection" was Grey. Despite the official authorization, the censor did hold up at least one of Plunkett's coded cables. Plunkett diaries, 5 June 1915. Vice Admiral Sir Henry Oliver, chief of the admiralty war staff, functioned as Balfour's private secretary.

13

A Private Diplomat

BOTH THE U.S. AND U.K. governments tolerated Plunkett's irregular diplomatic exchanges, in part because of the awkward state of normal diplomatic channels between the two nations. In Washington, D.C., Secretary of State William Jennings Bryan was too pacifistic to be of any help in dealing with the situation in Europe. On the other hand, Walter Hines Page, the American ambassador in London, was too pro-British to be seriously credited by some of his superiors in the United States. On the British side, Sir Cecil Spring-Rice, ambassador in Washington, was not physically well and was not particularly sympathetic to the United States anyway, making him a poor advocate for the British cause. House, as Wilson's eyes and ears, circulated among all the relevant parties, but he was also a man without a formal portfolio. Plunkett's private initiatives were thus fairly consistent with this rather chaotic situation.[1]

Nevertheless, this irregular situation troubled Plunkett, and in the summer of 1915 he wrote to House to explain his difficulties. He and House agreed that Balfour could be privy to their communications, but others in the British government could not be taken

[1] Plunkett referred to House as Wilson's *fidus Achates*, from Aeneas's "faithful companion" in Virgil's *Aeneid*. Plunkett's diaries, 11 Jan. 1915. Page loved everything British, and although he tried to observe the necessities of a neutral posture for the United States, he said, "A government can be neutral, but no *man* can be." Walter Hines Page to his brother, Henry A. Page, quoted in Hendrick, *Life and Letters of Walter H. Page*, vol. 1, 361. It is interesting that the British foreign secretary, Sir Edward Grey, did not entirely share the American view of Page as an anglophile. He said, "Page was of authentic British stock, but he came to London *absolutely and entirely American*" (italics added). Grey, *Twenty-Five Years*, vol. 2, 101.

"altogether in my [Plunkett's] confidence." Instead, some of their work would be that of "two agents having no official position." Plunkett concluded by saying that if he should be unable to serve House and the president, it would only be because the "peculiar" circumstances had "overmastered" his diplomacy. Under this arrangement, Plunkett sometimes went beyond giving advice to House and supplied specific information the latter requested. One example of such information was a complete list of British orders for munitions with U.S. firms in the period before June 1, 1915.[2]

After House's departure, Plunkett went to Balfour to ask him to get the British Press Bureau to prevent unfavorable press comments on American motives for or against intervention in the war. Of course, this sort of action was beyond the powers of the first lord, but Plunkett hoped that Balfour would take the matter to Asquith, the prime minister.[3]

An example of just how tangled the communications channels were occurred on June 9: the American Embassy in London asked Plunkett to tell House, who was then on board a ship bound for New York, that William Jennings Bryan had resigned and that his resignation had been accepted. Obviously, the embassy should have been able to send the message directly, but nevertheless Plunkett sent the radiogram to House on board the ship.[4]

Balfour must have expressed some question as to why he should be receiving advice on American policy from House via Plunkett, as Balfour and Plunkett drafted a letter for Plunkett to send to President Wilson regarding House's role in London. In this letter, Plunkett noted House's forthcoming absence from London, saying, "I have offered to be of any assistance in my power should misunderstandings arise in his absence." This letter predictably elicited what Plunkett called a "very nice" response from the president, who thanked Plunkett for helping House. In fact, the letter, which was clearly intended to be shown to others, was more than a "very nice" response: "It is fine to know you are so generously willing to play the

---

[2] Horace Plunkett to E. M. House, 8 June 1915, Plunkett collection, Library of Congress.

[3] Plunkett diaries, 7 June 1915.

[4] Bryan refused to sign Wilson's note to Germany regarding submarine warfare and instead resigned; Lansing, who was immediately appointed, then signed the note. Plunkett commented that the note would have been stronger if House had been in Washington. Plunkett diaries, 9 and 11 June 1915.

part you suggest and you may be sure that there would never be the slightest hesitation on my part about resorting to you in any possible contingency."[5] Plunkett showed this letter to Balfour, filling in between the lines, to assure the first lord that Wilson was firmly backing the Allies, although the Americans could not be more definite in writing because of German spies in the U.S. post office. By this exchange, the United States retained what the Central Intelligence Agency later called "plausible deniability" regarding Plunkett's assurances to Balfour. In case of a subsequent question, there was no paper trail implicating the president.[6]

The United States' insistence on "freedom of the seas" was a major source of friction with the British, and Plunkett asked Balfour to draft a paper to be forwarded to House that would protect British maritime interests. Plunkett then wrote to House asking that the Americans not deal with the issue of freedom of the seas until the Balfour paper arrived. Unfortunately, two months later, no letter had come from Balfour. The admiralty only acted after they learned that the Americans were about to issue new regulations. Plunkett then received the expected official letter and immediately cabled House, telling him to expect the letter and giving the gist of it as best he could, using his rather awkward code.[7]

While Plunkett and his British friends anxiously awaited a response from House, practical politics intervened regarding the U.S. cotton trade. When enforcing its blockade of Germany, the British delayed and sometimes intercepted shipments they considered to be destined for the Central Powers' war effort, and commercial interests in the United States pressured Wilson to have this practice halted. House sent a cable, which Plunkett said was "very urgent," warning that the United States would have to have some concession from the British in the matter of cotton sales, or Congress would embarrass the president.[8]

---

[5] Horace Plunkett to Woodrow Wilson, 4 June 1915, Link, ed., *Papers of Woodrow Wilson*, vol. 33, 338–39. House's diary entry of 24 June 1915 noted that the president had quoted Plunkett's praise of House "verbatim." Also, Woodrow Wilson to Horace Plunkett, 17 June 1915, Link, *Papers of Woodrow Wilson*, vol. 33, 414–15.

[6] Plunkett diaries, 10 July 1915.

[7] This code was held up at the censor, and Plunkett had to spend a day at the admiralty to get it sent. Plunkett diaries, 7 and 9 July 1916 and 16 Sept. 1916.

[8] Plunkett diaries, 20 July 1915. The House cable was sent the day before.

If the *Lusitania* affair had not been sufficient provocation to the United States, the German submarines provided yet another incident in August when the *Arabic* was torpedoed without notice, with the loss of four Americans (most of the passengers were saved). Plunkett now felt certain that Wilson would enter the war, but a month later, he again lost hope. His sense of foreboding further increased at the end of September 1915 when the British and French military commanders decided to try to resume the offensive all along the Western front. Instead of being elated, as he might well have been a year earlier, Plunkett dreaded the worst. "I . . . fear a calamity," he wrote in his diary.[9]

At the end of October, House wrote to Plunkett urging Plunkett to use the excuse of checking on business affairs to come over to America soon. Plunkett cabled back that he would come, but that he thought it advisable to take Balfour into his confidence, as their relationship was truly personal (House subsequently agreed). At a lunch alone with Balfour, Plunkett told the first lord that he might be able to see Wilson and asked what line he should take with the president. Balfour gave him "his entire confidence," according to Plunkett.[10]

At the end of this short meeting (Balfour could only give Plunkett an hour), Balfour agreed that Plunkett should frame another code they could use when communicating with each other. So it was that this private diplomat now served two governments, using different codes to keep his words secret both from the world and from his two masters. Recognizing the very irregular nature of his status for perhaps the first time, Plunkett asked Balfour to give him some official authorization, which he called "pseudo-official" and "only for safety's sake."[11]

Before departing for the United States, Plunkett met with Herbert Hoover, the future president of the United States, whom he had first seen the previous spring. Hoover was then running the huge and very successful American relief operation in Belgium and France, and he was making major waves on both sides of the battle

[9] Plunkett diaries, 19 Aug. 1915 and 25 Sept. 1915.

[10] House did not cable his permission regarding Balfour until 19 Nov., which may imply the need for some advance clearance in Washington. Plunkett diaries, 30 Oct. 1915 and 19 Nov. 1915.

[11] Plunkett diaries, 21 Nov. 1915.

line. He was not overly concerned with social niceties, with the result that Prime Minister Asquith called him impertinent, and Winston Churchill said he was a "son of a bitch."[12]

In Plunkett's first meeting with Hoover, the American was not very talkative, although he did comment that he was amazed at British naiveté regarding the strength of the German army. This time, Hoover had more to say, although what he said did not please Plunkett. Hoover delivered a lecture on the British conduct of the war and, in particular, their failure to place the country on a war footing. Plunkett recorded in his diary the following:

> His comments on the English at war was awful to listen to. We were simply *mad*. The extravagance of the war was a worse punishment than the Central Powers could devise. The 'blockade' had strengthened Germany (1) by forcing her to *economise* all round (2) by making every family curse England at every meal—hate a real power in war (3) by injuring us with neutral powers. It had at the same time demoralised England. We had entered into a wild extravagance of living. Generally, trade never so prosperous! Had the pinch of war been felt we should have won long ago. After the war we should have 15 years of desperate poverty. The rich classes wd be taxed 75% of their incomes as the means of working the revolution quietly! Chaos & disorder![13]

The circle of people who knew of the House/Plunkett relationship was perhaps larger than Plunkett thought, for Hoover, who was just back from the United States, confirmed to Plunkett that House urgently wanted to see him.

Plunkett sailed from Liverpool, and this time the ship, escorted by a destroyer, turned north along the Scottish coast before turning west into the Atlantic. Despite the supposed secrecy surrounding Plunkett's journey, no fewer than eight news photographers took his picture when he landed in New York on December 3. Another potential security breach was the fact that young Clifford Carver, private secretary of the American ambassador in London, was also there to greet the (supposed) businessman Plunkett.[14]

---

[12] Years later, Hoover said that his decision to help with Belgian aid was the worst mistake of his life, as otherwise he would never have been elected president. Richard Norton Smith, *An Uncommon Man*, 81, 86.

[13] Plunkett diaries, 22 Nov. 1915.

[14] Clifford Nickels Carver, who was born in 1891, was in contact with Plunkett several times in London. Plunkett diaries, 3 Dec. 1915.

When he met House, Plunkett got some bad news. House told him that Wilson could not go further than public opinion would support, and the United States was not coming into the war immediately, although the Americans were determined to "see the Allies through." Moreover, Wilson was beginning his reelection campaign, and this was a bad complication. House said the vote would be a sort of *Amerika über alles,* a declaration that the nation would not be dictated to by others.

House then brought up the subject of the American annoyance with the British ambassador. Poor Spring-Rice was the slave of his nerves, and those in the State Department, including the new secretary of state, Robert Lansing, found him impossible to negotiate with. House made the amazing suggestion—which he said enjoyed the support of both Wilson and Lansing—that Plunkett should be named ambassador. Remarkably, Plunkett's first thought on hearing this idea was that Washington, D.C., was the "most unhealthy" post in the diplomatic service, and he would surely collapse physically if he went there to live. Then, Plunkett pointed out to House that he had no wife, no great personal fortune to support the demands of that expensive post, and was not fluent in any foreign language.[15]

It is interesting that in his diary musings, Plunkett didn't mention the difficulty of getting the British government to appoint him to Washington. After all, he had been fired from the only British cabinet position he had held, in no small part because of his stubborn independence in political matters, and he had just emerged from a long fight with the government to secure a small grant for his Irish cooperative movement. It was simply not in the cards to suppose that the divided British government would name Horace Plunkett to represent the country in such a sensitive post as Washington then was.

So, it is just as well that Plunkett concluded for all of his *other* reasons that the ambassadorship didn't make sense and congratulated himself with the conclusion, "It is a delightfully American idea!" Even though he rejected consideration of the formal position, Plunkett's mere presence in Washington helped the U.S.

[15] In his diary entry of 14 Oct. 1915, House noted that he had suggested Plunkett as ambassador from the United Kingdom and Wilson had agreed.

relationship with the British embassy, for House noted that when Plunkett appeared, Spring-Rice seemed to be calmer.[16]

Plunkett admitted to himself that Spring-Rice was not helping the British cause, but otherwise, he did not seem to have encouraged American efforts to have the ambassador recalled. And House continued to think of ways to use Plunkett to improve relationships with the United Kingdom. At the end of 1916, House speculated with Wilson that if Edward Grey were to leave the foreign office and be replaced by Balfour, Plunkett would have "great influence" over British foreign policy toward the United States.[17]

In Washington, Plunkett met with Sir Cecil Spring-Rice at the British embassy and then went on to the State Department, where he saw Robert Lansing, the new secretary, and Frank Lyon Polk, a State Department lawyer. Polk lectured Plunkett on what he perceived was the mistake of the British blockade. Using a variant on Hoover's argument with Plunkett, Polk declared that the blockade actually aided Germany by preventing the Germans from bankrupting themselves by buying from the Americans. The next day, Plunkett returned for another visit with Lansing and Polk, which he did not elaborate on.[18]

Plunkett then went to the Department of Agriculture to discuss agricultural cooperation, which was part of his "cover" story for the trip to the United States, and then he met other cabinet officials, where the conversation always included the subject of American politics. He saw Interior Secretary Franklin Knight Lane twice to discuss irrigation as well as politics. Lane felt the United States could not help the Allies in the war, and he was pessimistic about Wilson's reelection chances. At the Treasury Department, Plunkett met Secretary William Gibbs McAdoo, who was also Wilson's

---

[16] Plunkett's musings about the ambassadorship prepared him to respond when the subject was next mentioned to him. Early in 1916, Charles R. Crane, a wealthy contributor to Wilson's 1912 campaign, brought Plunkett an autographed photograph of Wilson and said that the president wanted Plunkett to be the British ambassador. This time, Horace did not hesitate but declared that the job would kill him in a month. Plunkett diaries, 3 and 10 Dec. 1915 and 13 Jan. 1916.

[17] E. M. House to Woodrow Wilson, 10 Dec. 1916, Link, *Papers of Woodrow Wilson*, vol. 40, 212. Balfour was subsequently appointed to the foreign office in December 1916 by Lloyd George, but by this time, Plunkett was seriously ill and House had another British informant, with his own private code.

[18] Plunkett diaries, 8 Dec. 1915.

son-in-law. McAdoo talked volubly, but without saying anything substantive, about the U.S. position on the war in Europe; at the end of the interview, he even said that the possibility of war with England could not be ignored.[19]

At the Navy Department, Plunkett saw Assistant Secretary Franklin Delano Roosevelt again to discuss the knotty question of whether a merchant ship could pass through the blockade if it had a defensive gun mounted on board. Plunkett volunteered to try to work with Arthur Balfour to propose an acceptable answer to the question. His crowded interview schedule also included meetings with Sydney Brooks of the *London Times* and *Daily Mail* and with Walter Lippmann of the *New Republic*, which had become pro-intervention after the sinking of the *Lusitania*.[20]

The interview with Sydney Brooks was a part of Plunkett's work regarding wartime propaganda. He scolded Brooks for "unwisely" arguing that British censorship had magnified the defeats of the Allies. Later, Plunkett learned that the Germans had offered an "enormous bribe" to a New York journalist to support their cause, and he hoped to organize a syndicate to mount a pro-Ally American propaganda offensive in the Midwest; he even selected the man to run the operation. Later on, when he was back home, Plunkett tried to dampen criticism of America in the British press, and when House complained about some *Punch* cartoons critical of Wilson, Plunkett passed the complaint on to the magazine. In Edinburgh, Plunkett "fully instructed" the editor of the *Scotsman* on the "correct" attitude toward America. Although Lord Kitchener at the War Office was delighted with Plunkett's publicity plans to "buy" the American press, the Foreign Office was not pleased. When Plunkett got the approval of Balfour and Lord Robert Cecil (minister of blockade) to send a cable making the case for American inter-

[19] Plunkett tried to get a sense of U.S. policy from McAdoo by suggesting that the United States would need to obtain satisfaction for the *Lusitania* sinking, and McAdoo replied, "Ah! There's the trouble. The situation is grave." Then Plunkett asked about the Ancona sinking the month before, and McAdoo said, "Yes, that too!" Plunkett diaries, 10 and 11 Dec. 1915. MacAdoo married Eleanor Randolph Wilson in Washington, D.C., on 7 May 1914.

[20] Plunkett diaries, 9 Dec. 1915. Walter Lippmann, who was 26 when Plunkett met him, had joined with Herbert Croly the year before to establish the *New Republic*. In 1917, he was appointed assistant to Secretary of War Newton Baker, and he was secretary to the Inquiry "brain trust" that helped to draft Wilson's Fourteen Points.

vention, Lord Grey blocked it until after Plunkett personally met with him. Plunkett complained to his diary that the government publicity machine was too entrenched to be scrapped.[21]

Leaving Washington, Plunkett went to Omaha, where he really did have a business problem he needed to deal with. There was a potential British tax problem regarding his U.S. properties, and he also needed to reorganize his estate to make probate simpler. The 1914 change in the British tax law potentially subjected his worldwide earnings to British income tax at rates as high as 50%, whether the money was repatriated to the United Kingdom or not. On the recommendation of his auditor, T. C. Cannon, Plunkett organized the Nebraska and Wyoming Investment Company (hereafter the Nebraska Company) to take over the Nebraska and Wyoming properties. The company was capitalized at $800,000, and Plunkett initially held all of the stock.[22]

While in Omaha, Plunkett continued to sample American public opinion regarding the war in Europe. He was concerned about the large German population in the Midwest, but the businessmen he spoke with did not think that factor would affect U.S. policy on the subject. Nevertheless, the cross-currents in American opinion were apparent later when he made his periodic visit to Battle Creek. At Dr. Kellogg's request, Plunkett gave a talk on the European war, and this time a German member of the audience accused him of breeding hate.

---

[21] Brooks, who was in the United States representing the *London Times* and *Daily Mail,* blamed the failure of the Allies' loan in Chicago on British censorship, which made Americans skeptical of Allied prospects in the war. *New York Times,* 26 Nov. 1915; and Plunkett diaries, 8 Dec. 1915. The "enormous bribe" was supposedly offered to Edward Marshall, a New York journalist who was later to be chief executive of Plunkett's publicity effort. Plunkett personally financed the planning for the publicity campaign and advanced £500 for the work, although he expected some reimbursement from the British government. Horace Plunkett to Karl Walter, 16 Aug. 1917, Plunkett collection, Library of Congress; and Plunkett diaries, 16, 19, and 28 Feb. 1916; 1, 2, 21–25 March 1916. Sir Edward Grey was elevated to the peerage as Viscount Grey of Falloden in 1916 and thereafter was addressed as Lord Grey.

[22] Plunkett diaries, 13 Dec. 1915; and Horace Plunkett to Frank C. Bosler, 5 Aug. 1915, Bosler collection. The Nebraska company was organized 15 Dec. 1915, and real estate valued at $644,821.33, subject to mortgages of $215,000, was conveyed to the corporation on 18 Dec. H. F. Heard to Conrad Young, 3 June 1933, NSHS collection. T. C. Cannon complained that Conrad Young had convinced Plunkett to transfer the Wyoming companies to the Nebraska company, saying that Young wanted to liquidate those holdings. T. C. Cannon to S. Sharpe Huston, 17 Feb. 1917, Bosler collection. The new corporation issued both common and 6 percent preferred stock.

After trips to Omaha, Cheyenne, and Battle Creek, Plunkett returned to New York, where he saw House before the latter left for London. Then Plunkett went back to Washington, D.C., to see the British ambassador again. Spring-Rice told him that the U.S. crisis over the *Lusitania* and *Ancona* would probably be settled diplomatically by Wilson. Plunkett took this to mean that the United States would continue its neutral position, and the war would go on. On January 13, he sailed for Liverpool.[23]

On the day after his arrival in England, Plunkett lunched with Balfour, and the next day Balfour showed him cables from Spring-Rice regarding the strained relations with the United States. Later that day, Plunkett went to see Sir Edward Grey, who seemed not to understand the United States at all. Plunkett wrote that Grey was lacking in imagination, saying, "He has a wonderful power of clear expression of such thoughts as he develops."[24]

At a meeting in London, House asked Plunkett, "Why won't your government give us the crumb which will enable us to give them the loaf?" but when Plunkett probed the metaphors, House could not define either the crumb or the loaf. House told him that he was now carrying a letter in which the president had given him absolute authority as spokesman. Plunkett again argued that Wilson should clearly support the Allies.[25]

Plunkett's diplomatic initiatives were interrupted by the Easter uprising of 1916 in Ireland, by further health problems, and by the endless saga of the Diamond Company. The Easter uprising profoundly shocked Plunkett, and he was actually fired on in the streets of Dublin (although by government guards, not by the Irish rebels). Then in the summer of 1916, he suffered a bad X-ray burn,

[23] During the train trip between Cheyenne and Omaha, Plunkett mused about his first trip to Wyoming Territory in 1879. In his diary he wrote, "I don't think the ten years in the West were wholly wasted though doubtless they might have been better used." Plunkett diaries, 14–19, 23, and 27–30 Dec. 1915; 13 Jan. 1916.

[24] Plunkett diaries, 23 and 24 Jan. 1916. Grey was noted for evasive answers, but he did have a way with words. On the evening of 3 Aug. 1914, the day Grey had thrown down the gauntlet to Germany over the invasion of Belgium, Grey watched the street lamps being lit. He then uttered his most famous line, saying, "The lamps are going out all over Europe; we shall not see them lit again in our lifetime." Tuchman, *Guns of August*, 116, 123.

[25] House told Plunkett that Bryan resigned as secretary of state because House suggested Wilson send a strong note to Germany, whereupon Bryan said that House was the secretary of state.

which created pain that could not be alleviated even with cocaine. Hence, he went to Kilteragh to recuperate in a bed on the roof, where he was attended night and day by nurses. By the end of August, the doctor thought Plunkett could recover in two months, but he was still taking morphine in November. During this period he could do little more than dictate a few letters.

At the beginning of 1916, Plunkett gave Bosler a written offer proposing Bosler buy Plunkett's Diamond Company stock at a price of $60,000, provided his notes were also paid off in "cash, or as good as cash." Bosler did not immediately take up the idea, expressing "regret" that Plunkett wanted to give up the property; however, in the margin of Plunkett's letter, he calculated that a $60,000 price for the stock would give Plunkett a 33 percent gain.[26]

Finally, in the fall of 1916, the Diamond Company sold the mortgage bonds, although on less favorable terms than Bosler had hoped to obtain. Still, there was not enough money, and Bosler paid Plunkett only the interest arrearages on his notes, leaving outstanding the $60,000 as well as the $20,000 note at the Omaha bank that Plunkett had guaranteed. When he learned what Bosler had done, Plunkett was still in bed, trying to recover from the X-ray burn, and he was not in a good mood. In a letter dictated to a stenographer, Plunkett complained that Bosler was using the available money for "innumerable" claims of lower priority than his.[27]

Now that he had transferred his Diamond Company shares to the Nebraska holding company, Plunkett told Bosler to contact Conrad Young on "all matters." Despite this formal washing of his hands on Diamond Company matters, Plunkett continued to write to Bosler venting unhappiness and accusing Bosler of assuming that the Plunkett notes were "mere formalities, only to be acted upon when the Diamond Co. could meet the obligation without inconvenience." The final deal to buy out Plunkett's share of the Diamond Company was far less favorable than Plunkett's earlier offer to Bosler, giving no immediate cash and extending the notes for periods reaching out to the end of 1920. Bosler did not meet

[26] Plunkett owned 452 shares of the Diamond Co., which had been valued at $45,200 at the time the company was organized. Plunkett's offer was to expire on 1 April. Plunkett diaries, 8 Jan. 1916; Frank C. Bosler to Horace Plunkett, 4 Jan. 1916; and Horace Plunkett to Frank C. Bosler, 12 Jan. 1916, Bosler collection.

[27] Frank C. Bosler to Horace Plunkett, 1 Aug. 1916, Bosler collection.

even these reduced terms, and payments on the notes continued after both men died.[28]

At the end of May, President Wilson made an important speech urging the right of people to choose their government, but unfortunately he also said that America was not concerned with the causes and objects of the war. This speech was generally applauded in the United States but not in Europe, on either side of the conflict. The Germans thought it was a reelection gambit, and Lord Grey told House that the president's objectives could be achieved only after a peace won by the Allies.

As the summer wore on, Anglo-American relations declined to a level not seen since the burning of Washington in 1814, according to one historian. The Allies began seizing American mail, the British prepared a blacklist of American firms suspected of dealing with the Central Powers, and Secretary Lansing fretted that the United States might have to side with the Germans. A clear barometer of the diplomatic climate was the fact that House and Grey corresponded only sporadically. To further complicate matters for the United States, Pancho Villa led an incursion into New Mexico, and Wilson sent General Pershing into Mexico to apprehend Villa, infuriating public opinion in Latin America and making war with Mexico a possibility.[29]

Finally, the doctor gave Plunkett permission to sail for the United States at the beginning of December 1916. Just before departing, he learned with satisfaction that Arthur Balfour had been appointed foreign secretary, replacing Lord Grey. In New York, House told Plunkett that the idea of U.S. intervention was no longer "practical politics," and that the United States had nothing to fear from Germany and Japan, although the United Kingdom could imperil the United States. Before he could try to correct this view with House and Wilson, Plunkett had to go to Battle Creek for health reasons, and he remained there until the beginning of March 1917.

---

[28] Plunkett had to accept a further 20 percent discount in the value of his stock. The notes to purchase the 452 shares of stock were dated 6 Jan. 1917 and payable in October of 1921 and 1922. The existing notes totaling $75,000 were extended to fall due in the years 1917–20. Horace Plunkett to Frank C. Bosler, 24 Aug. 1916, 26 and 26 (second letter) Oct. 1916, and 31 Dec. 1916; and Frank C. Bosler to Conrad Young, 13 Jan. 1917, Bosler collection.

[29] Knock, *To End All Wars,* 76–81.

After leaving Battle Creek, Plunkett optimistically contemplated a trip to Chicago, Madison, Kansas City, and Omaha, but the train journey to Chicago opened up the partially healed areas on his body, and after suffering considerable pain for three days, he entered Presbyterian Hospital. He placed himself in the care of Dr. Arthur D. Bevan, a distinguished surgeon who agreed to perform the surgery without the use of a colostomy. The surgery was successful, and Plunkett was able to move from the hospital to a hotel by the middle of April. In the midst of this medical regimen, the stream of letters to and from Plunkett continued unabated, including a letter from House offering him a reserved seat in the president's stand at the inauguration. A week later, Plunkett was in Washington, where he wrote, "The joy of being back in the middle of big affairs is inexpressible."[30]

The diplomatic landscape had changed radically during Plunkett's medical interregnum, for German belligerency and the Zimmermann telegram finally pushed Wilson to declare a state of war with the German Empire on April 2. Plunkett's role was diminished, only partly because of his forced convalescence. At the end of 1916, Sir William Wiseman, the new head of British intelligence in the America, came to the United States to meet House. The two men became close friends, and Wiseman soon became the conduit between House and Balfour at the foreign office, using a private code, much as Plunkett and Balfour had done earlier. Plunkett was aware of this new relationship but did not comment on it.[31]

The new reality created solidarity between Washington and London, and there was less need for informal intermediaries. Balfour came to Washington for a visit with the U.S. authorities, and Plunkett met him there; although Plunkett's reception was cordial, his recommendations were largely ignored. On the first of May he set sail for Liverpool on the *Adriatic*, and for the trip he was given a waterproof suit that would also keep him warm if he were to find himself in the cold Atlantic waters. The strongest manifestation of

---

[30] Plunkett diaries, 6 March– 21 April 1917. Dr. Bevan was president of the American Medical Association in the years 1918 to 1919.

[31] Hodgson, *Woodrow Wilson's Right Hand*, 131–32. In the fall of 1917, Wiseman sent House two of Plunkett's confidential reports on the Irish Convention without informing Plunkett, who was then authorized by Balfour to forward the subsequent reports as well. Horace Plunkett to House, 28 Sept. 1917, in Seymour, *Intimate Papers of Colonel House*, vol. 3, 76–78.

the new dynamics in the war came off the coast of Ireland, where the liner was met by its naval escort: this time the escort was an American destroyer.[32]

Back home, Plunkett's attention was entirely taken up by the Irish Question. He was named chairman of the convention to recommend a government for that unhappy island, and both the King and President Wilson asked him to keep them informed on this work. Unfortunately, the task he was charged with defied a negotiated settlement, and the British government did not relish the prospect of American involvement in it. At the end of 1918, when he tried to have a government official deliver a letter on the Irish situation to House (who was then in Paris), the British government blocked the letter.[33]

In Washington, D.C., early in 1919, Plunkett's reception was also cool. Before Plunkett arrived in Washington, President Wilson announced publicly that he would not interfere in the Irish Question, and Wilson's private secretary sent word that getting an appointment with the president would be "impossible." Plunkett then spoke to Lansing's private secretary, who opened a conduit to Wilson through Frank L. Polk, who was then acting secretary of state. By this means, Plunkett's sealed letter reached the president.[34]

In Chicago, Plunkett spoke to Irish Americans, and the *Boston Globe* headlined the event: "Plunkett's Talk Arouses Storm." There were now opportunities for interviews with the *New York World* and the *Christian Science Monitor*. Finally, Plunkett was able to see Frank Polk, and they had a long talk about the Irish Question. Polk emphasized the need for an early settlement because of the effect on the U.S. attitude toward the League of Nations.[35]

Plunkett's record of amateur diplomacy is an interesting example of the way foreign affairs are sometimes conducted outside official channels. An out-of-favor former government official, Plunkett

---

[32] Plunkett diaries, 30 April 1917 and 1 and 9 May 1917.

[33] Plunkett asked Assistant Undersecretary William G. Tyrrell to deliver the letter. Plunkett diaries, 19 Dec. 1918.

[34] Plunkett diaries, 28 Feb. 1919. Wilson responded to Plunkett's letter on 26 March from Paris, in a letter that did not reach Plunkett until 1 May. Wilson was gracious but said only, "I am very glad indeed to have your guidance in a most perplexing matter."

[35] Plunkett diaries, 11 and 15 March 1919.

operated in a sort of shadow land of diplomacy, at times trying to influence the British government, and at other times seeking changes in American policy. His use of private codes with both governments was completely innocent, but for a private person to do so, albeit with the complicity of the British Admiralty, is at least unusual.

We cannot know the true extent to which Plunkett was used by, or merely tolerated by, his diplomatic contacts. With the possible exception of Arthur Balfour, British government officials seem generally to have confined their conversations with Plunkett to matters that were already known by the public or would shortly be so. House *appeared* to take Plunkett deeper into his confidences, but it is never possible to know whether the ideas he floated were really Wilson's, or whether they were only pure House (even though he honestly may have believed he could sell them to Wilson).

On his side, Plunkett trusted House implicitly, and at one time he wrote, "I am absolutely satisfied he is . . . the embodiment of truth." To demonstrate his authority, House was carrying a letter signed by Wilson that named him as "spokesman" for the president, and he told Plunkett that he had "absolute" authority in that role. Plunkett did not always tell House everything he had learned from his British sources (or from his American sources, either, for that matter), but he had no doubt that House could be trusted with any confidence he provided to the Texan.[36]

Regarding the effectiveness of Plunkett's relationship with House, we must acknowledge that Plunkett was only one of the many contacts House used as he spread a wide net to collect information to feed back to Wilson. Thus, it is possible that almost nothing that he learned from Plunkett was unique. Nevertheless, House trusted Plunkett enough to use him as an important source to check what he was hearing from others.

---

[36] The first ellipsis in the quoted portion is the word "not," which is crossed out in the original. From the context, it is clear that Plunkett started to write "not a liar," but substituted the stronger version in the quote. Plunkett diaries, 7 Dec. 1915 and 11 Feb. 1916.

14

The Last Years

PLUNKETT WAS TREATED to a preview of his own obituary in the spring of 1920 after he refused to give a New York reporter an interview, and the reporter then gave out the news that Plunkett was dead. By the time he got to Battle Creek, his London solicitors were already asking what would become of his remains. Although this news of his death was a hoax—a dozen years premature—Horace's life was on a downward slope.[1]

Early in 1919, he weighed only 114 pounds when he went back to Battle Creek to get help in overcoming his morphine addiction, which arose from taking the drug to permit him to work and to sleep. Unfortunately, Dr. Kellogg was not there, having gone to Florida for the winter because of his own health problems. In typical fashion, Plunkett used the available time to send Kellogg the benefit of his advice on reorganizing the sanitarium. He wrote a letter, endorsed by three of the Battle Creek doctors, in which he suggested establishing a Race Betterment Movement that would be centered on the Battle Creek sanitarium.[2]

---

[1] When Plunkett returned to Ireland, Thomas Power O'Connor told him he had received £15 to write Plunkett's obituary for the *Daily Telegraph*. Although the obituary did not appear, O'Connor said he would save it for Plunkett's "next death." Plunkett diaries, 1 and 2 Jan. 1920.

[2] Plunkett's letter was "strongly" endorsed by Dr. (Lt. Col.) James T. Case, Dr. William Riley, and British surgeon Dr. J. H. Patterson. Plunkett diaries, 4, 12, 22, and 23 Feb. 1919. At the Race Betterment Congress held at the Panama–Pacific Exposition in 1915, Dr. Kellogg advocated establishing a eugenic register that would create a pedigree of proper breeding pairs. *San Francisco Chronicle*, 8 Aug. 1915. Kellogg was the founder of the Race Betterment Foundation. The Battle Creek sanitarium fell on hard times and went into receivership in 1933.

Plunkett's suggestion was not an original idea but, rather, built on Kellogg's involvement with the eugenics movement in the United States extending at least as far back as 1906. In that year, Kellogg founded the Race Betterment Foundation, and in 1914, he held the first Race Betterment Conference in Battle Creek. The object of this work was to create a super race. "We have wonderful new races of horses, cows and pigs," Kellogg said. "Why should we not have a new and improved race of men?" His vision was of a race of "Human Thoroughbreds" created from the white races of Europe. These ideas even received support from Theodore Roosevelt, who wrote in 1913, "I agree . . . that society has no business to permit degenerates to reproduce their kind. . . . The inescapable duty of the good citizen of the right type, is to leave his or her blood behind him in the world."[3]

Kellogg's race betterment concept consisted of two strategies. His "euthenics" program of personal and public hygiene included vegetarianism, electric baths, enemas, bran consumption, and public health surveys. He coupled this program with a proposal for a national eugenics register to identify mental deficiencies in the population (which could then be dealt with by sterilization). By the 1920s, these ideas had fallen in disrepute among prominent eugenicists. It is unknown precisely what Plunkett's specific contribution to this subject was, but fortunately he apparently did not expend further effort on it.[4]

Plunkett was at Battle Creek at the beginning of 1920, the beginning of 1921, and again at the end of that year, when he stayed until late January 1922. In the fall of 1923, Dr. Riley wrote to him, saying Plunkett could not return to Battle Creek unless he gave up his use of morphine. However, this attitude later changed, and at the beginning of 1924 Plunkett spent a month and a half at Battle Creek, still trying to cure his addiction.[5]

At the beginning of 1921, Plunkett had to deal with the problems of Daisy Fingall's son Gerald, who was in Kansas City supporting himself in Prohibition America by smuggling whiskey. Although Plunkett thought he had "saved" the boy after their conversation, at the end of the year more trouble arose, causing Plunkett to remem-

[3] Black, *War against the Weak*, 88, 99.
[4] Stern, *Eugenic Nation*, 53.
[5] Plunkett diaries, 1 and 26 Jan. 1922, 23 Aug. 1923, 4 Jan. 1924, and 13 Feb. 1924.

ber a quotation of his father's that fit the situation: "You cannot make an empty sack stand."[6]

As the Irish question heated up, it was more difficult for Plunkett to remain on cordial terms with both sides of that question. His trip to the United States at the beginning of 1920 was at least in part to secure support there for a propaganda service to promote his dominion plan for Ireland, and the powerful Sinn Fein lobby in the United States bitterly opposed him. For the first time, Irish businessmen did not want to be seen with Plunkett, and even Bourke Cockran was no longer friendly.[7]

Early in 1922, he spoke to a capacity crowd in New York's Town Hall and was heckled afterward. A year later, when he again spoke at Town Hall, he had a motorcycle escort in front and behind him and had to be "smuggled" out of the hall through a back door to avoid a crowd of "wild" women. Unfortunately, these events were merely an American reflection of the more violent forces afoot in Ireland.[8]

At the end of January 1923, Plunkett was in Wisconsin when he received word that the servants at Kilteragh had been turned out and the house blown up. He lost all of his personal belongings, except for the few photographs at Plunkett House and some "odds & ends" at the London flat. Plunkett then moved his residence to England, where he bought Crest House. Afterward, he seldom visited Ireland, and he lived out the rest of his life on the outskirts of London as a sort of exile.[9]

At the beginning of 1923, the Senate Committee on Agriculture

[6] Plunkett diaries, 7 Jan.–10 Feb. 1921 and 30–31 Dec. 1921. Gerald Plunkett first attracted official attention when Edward VII, who was in Ireland with the Queen, said to Daisy Fingall, "And there is that horrid little boy of yours treading on the Queen's train." In World War I, Gerald falsified his age and enlisted as a driver, and after the war he was a driver for Lord John French, who was Viceroy of Ireland from 1918 to 1921. Hinkson, *Seventy Years Young,* 101, 364. Both Gerald and his older brother, Oliver, died without male issue; hence, when Oliver died in 1984, the earldom of Fingall became extinct.

[7] In Ireland, Plunkett's cooperative work suffered from his identification with the dominion issue, and in the fall of 1920 some twenty creameries were destroyed. Plunkett said that Sinn Fein was a powerful factor in blocking U.S. ratification of the Versailles Treaty. Plunkett diaries, 29 Jan. 1920, 4 and 13 Feb. 1920, 19 Aug. 1920, and 1920 summary.

[8] Ibid., 9 Feb. 1923.

[9] Obviously, at least some of Plunkett's diaries survived the bombing, and after he moved to Crest House, the furniture salvaged from Kilteragh proved more than the smaller abode could accommodate. In short, it seems the story of the Kilteragh bombing was not fully explained. Ibid., 28 March 1921, 20 May 1921, 26 March 1922, 22 April 1922, 10 July 1922, 15 Nov. 1922, and 22 Dec. 1922.

and Forestry held hearings on the cooperative movement in the United States, and Plunkett was invited to testify. Plunkett began by sketching the history of the cooperative movement in Ireland, but he had not gone far when some of the senators bluntly told him they wanted to hear about present-day matters, not history. After these preliminaries, Plunkett plunged into the real controversy surrounding the cooperative movement in the United States. He said that one had to be careful about urging cooperation, because businessmen and some politicians did not want the farmers to organize. Plunkett liked the use of cooperatives to organize farmers, because each member had a single vote, and there were no controlling shareholders.

He argued that organizing farmers would not increase food prices, and he suggested that the best way to accomplish organization was to teach city folk the value of a prosperous rural community, with rural residents who would be able to buy goods and services from the city. At the conclusion of his testimony, Senator Kendrick from Wyoming endorsed his comments, noting that farmers and stockmen had to buy their supplies in a price-controlled market but had no influence on the price at which they sold their products.[10]

As previously mentioned, Plunkett had incorporated his holdings in Nebraska, with himself initially as sole shareholder. Later, he gave Young, Cannon, and Doherty shares in the Nebraska Company in lieu of a cash bonus and as an incentive and reward for their work in managing the American properties. One can infer that Young and Doherty together received at least 2,000 shares, which would imply a notional value of $200,000, making a very generous settlement with them. Almost immediately, Plunkett began to fret that the depreciation in the value of the Wheatland companies made the incentive smaller than he had intended, and he promised them an adjustment. For a number of years, cash dividends were paid to the smaller shareholders of the company, but Plunkett's dividends were simply credited to his trust account and not paid in cash, in the vain hope to avoid U.K. taxes.[11]

---

[10] U.S. Senate, *Hearing Relative to Cooperative Agricultural Associations.*

[11] Conrad Young and his sister also received 70 shares of the Nebraska Company. T. C. Cannon received preferred shares, perhaps to a value of $5,000. At the beginning of 1921, Plunkett gave 100 shares of the Nebraska company to Ralph Stuart-Wortley as additional compensation for acting as his financial representative in New York. (*continued, next page*)

Omaha and South Omaha were growing, and the Nebraska Company prospered, with profits of $19,000–$23,000 in the years 1923–25 and a corporate surplus of over $52,000 by the end of 1925. Starting in 1922, Plunkett began taking his dividends in cash, and after the income from his English properties declined sharply in the spring of 1925, he drew more heavily on his American investments. Young sold some bonds owned by the Nebraska Company, and these proceeds, together with principal collections on the Diamond Company and Wyoming Industrial notes, permitted him to send £5,000 to Plunkett.[12]

The rosy profits in Omaha came to an end in the fall of 1926 when the agricultural depression in the American West reached Omaha, which, Conrad advised Plunkett, was "largely overbuilt." Nevertheless, this financial setback did not disturb Plunkett, for he had become very comfortable with Young's handling of his affairs. In the fall of 1930, he wrote to Young, "You never have to consult me on business affairs as your judgment and information are in every respect superior to my own." Thirty years earlier, it would have been impossible for him to write such a letter. Just before his death in 1932, Plunkett was contemplating making an additional gift of some 20 to 25 percent of the Nebraska Company shares to Young. Although he did not make this gift, Plunkett advised his executor, Gerald Heard, to pay Young a commission on the liquidation of the American investments.[13]

Before Plunkett's death in 1932, Conrad Young prepared a valuation of the Omaha real estate, which apparently then consisted of only five properties: three were developed, and two were vacant lots. The cost of the properties and improvements was over

---

(*continued from previous page*) He also gave 100 shares to Karl Walter, who was at that time expected to be executor under Plunkett's will. Horace Plunkett to Conrad Young, 11 March 1918, 22 May 1918, 23 Dec. 1918, and 10 Nov. 1931, NSHS collection. After Plunkett's death, the U.K. tax authorities required his estate to pay taxes on his dividends from the Nebraska Company.

[12] The net income of the Nebraska Company was $21,153.97 in 1923, $22,976.96 in 1924, and $19,152.23 in 1925; surplus at the end of the last year was $52,043.25, and a 4 percent dividend on common was to be paid for that year. The Nebraska Company had some $73,000 in marketable securities and owed Plunkett over $61,000, consisting of collections on the Wyoming companies' notes, plus accrued dividends and interest. Conrad Young to Horace Plunkett, 9 May 1925, 25 July 1925, and 4 Dec. 1925, Plunkett collection, Library of Congress; and Plunkett diaries, 27 Jan. 1922.

[13] Horace Plunkett to Conrad Young, 8, 24, and 29 Oct. 1926 and 26 Jan. 1932, NSHS collection.

$530,000, and after deducting nearly $58,000 of depreciation, the net book value was $473,000. Although book values had previously been written down, they were still much higher than the current market, which Young thought was about $297,000.[14]

Although Nebraska real estate failed to make Plunkett a millionaire, as he sometimes dreamed, it still gave him the largest asset to pass on to his beneficiaries. In part, this positive result was possible because he rightly predicted rapid growth for Omaha, but it was also a product of patience and of good management in the Omaha office. Under the circumstances, it was only fair that these managers should be well compensated for their services.

Although Wyoming Development had assets in the form of unsold lands, sales did not proceed rapidly enough to offset the cost of running the company, and in 1924, the net loss was over $18,000, but the Industrial Company had a small profit. The following year, Wyoming Development lost over $16,000, and its liquid assets were rapidly being depleted. In 1926, the loss was still $13,000.[15]

The Wheatland Project ultimately became a magnificent farming community, but for many years it returned nothing to its promoters. The New York shareholders of the Wheatland companies were unwilling to have their investment tied up any longer and demanded that the companies be liquidated. Frank Sturgis, who had replaced his brother Tom on the shareholder list, even threatened to throw the companies into bankruptcy. Conrad Young, who also represented these shareholders in the West, counseled patience, and Sturgis then suggested a "friendly" auction of the unsold properties. Plunkett fretted that the project would be liquidated faster than Young and Cannon thought prudent, and when Robert Carey was elected to the U.S. Senate in 1930, Plunkett also worried he would have less time to devote to the Wheatland Project. Young was still caring for the investments in the two corporations when Plunkett died in 1932, and it was only afterward that his estate received any significant return.[16]

---

[14] Young's listing was appended to the end of Plunkett's 1930 diary.

[15] Conrad Young to Horace Plunkett, 19 May 1925 and 13 May 1927; and T. C. Cannon to Horace Plunkett, 26 April 1926, Plunkett collection, Library of Congress.

[16] In the fall of 1940, Young became disillusioned with Carey's handling of the project liquidation and contemplated selling the shares. Horace Plunkett to Conrad Young, 29 Oct. 1926 and 31 Oct. 1930; and H. F. Heard to Conrad Young, 18 Sept. 1940, NSHS collection. Also, Conrad Young to Horace Plunkett, 6 June 1925; and F. K. Sturgis to Conrad Young, 15 Oct. 1926, Plunkett collection, Library of Congress.

In the spring of 1927, Plunkett lost the man who handled his American securities investments when Ralph Stuart-Wortley died in New York. Stuart-Wortley, who was one of those young British aristocrats in the Powder River Basin in the 1880s, went to the East Coast after he left Wyoming. He finally settled in New York, and when he died he had a seat on the New York Stock Exchange. For a number of years, he had handled Plunkett's portfolio of American securities, and his death meant the loss of a trusted friend from Plunkett's American interests.[17]

In the spring of 1930, there was a final footnote to Plunkett's ranching experiences in Wyoming. Some forty years after he left the range in Wyoming, Plunkett had the sad duty of winding up the Ranchman's Association, a group of British gentlemen who tried their hand at ranching in the American West. Once numbering some 300 members, the association dwindled to the point that a circular letter to the membership elicited only two replies, one of which was from Boughton. Plunkett could not even find anyone to discuss what to do with the £350 the association still had.[18]

As mentioned earlier, the executor of Plunkett's estate was Henry Fitz-Gerald Heard. Plunkett first met Heard, who also exerted a profound influence over Horace in his declining years, near the end of 1919 when Heard was 30. Plunkett invited Heard to spend a few days at Kilteragh "to see whether we might suit each other," and the next day he declared that he liked Heard.[19]

The son of an Anglo-Irish clergyman, Gerald Heard was certainly the strangest man Plunkett ever chose as an associate and perhaps also the most brilliant. Flamboyant in appearance, he might appear for dinner wearing a leather jacket with leopard-skin

[17] Ralph Stuart-Wortley died in New York on 2 March 1927. *New York Times*, 3 March 1927. When he came to Wyoming, Stuart-Wortley was the younger son of a younger son, with long odds against succeeding to the earldom. However, sixty years after his death, his grandson, Richard Alan Stuart-Wortley, who lived in Connecticut, did succeed to the title in 1987.

[18] Plunkett wanted to give the fund to his foundation but could not get anyone to discuss the matter with him. Plunkett diaries, 14 Feb. 1930. Another sad ranch reminder arose when Erskine Booth's widow wrote to say she had letters from Plunkett dated in April 1911 in which he admitted owing Booth money. Plunkett noted that "beyond all question" he had paid the sum, but the proof was in the ashes of Kilteragh. Horace had resolved to make an ex gratia payment to her, but then the widow wrote again to say she had found proof that Plunkett had paid the £400 debt. Plunkett diaries, 2, 6, and 8 Feb. 1932.

[19] Plunkett diaries, 6 and 7 Dec. 1919.

collar and pointed purple suede shoes. He wrote over 35 books and could discourse at length about philosophy, psychology, and history. He has been described as having the hypnotic power of many evangelists. Plunkett was not the only person who was fascinated by Heard, for it has been said that people dropped their voices when they spoke of him, "as if he were Jesus Christ."[20]

Nevertheless, the euphoric beginning of Plunkett's association with Heard later deteriorated badly, and by 1927 Heard did not even tell Plunkett when he invited his young friends to the house, for the night or for several days. Heard was then devoting less than half his time to Plunkett's work, but Horace did not criticize his behavior, fearing that Heard would leave him. Nor did Plunkett dare to hire anyone else to help with the backlog of clerical work, as that also would undoubtedly cause Heard to leave.[21]

At the end of 1927, following a nasty quarrel, Heard abruptly left permanent employment with Plunkett, although he continued to return to look in on Horace nearly every week for the rest of the older man's life. For Horace, this began a somewhat pitiful period of depression, during which he successively hired and fired four "useless" private secretaries. Despite the unpleasant episodes that marred their relationship, Plunkett still counted Heard as a friend and never lowered his opinion of Heard's character and intelligence.[22]

When Plunkett's health began failing rapidly, Heard helped him to draft his final will, which was dated July 16, 1931. Unable to make the trip across the Atlantic to confer personally with Young, Plunkett summoned Young to come over to England. Young arrived there on July 27, 1931, where he met Gerald Heard for the first time. To Plunkett's delight, the two men seemed to like each other. Unfortunately, Plunkett was so weak that he slept through half of their meeting together. (Ironically, although Plunkett had been

---

[20] Ibid., 14 Dec. 1919, 16 Jan. 1927, and 21 Feb. 1927. Heard was a BBC commentator on scientific subjects in the years 1930–34. The "Jesus Christ" quote is by Leo Charlton in Furbank, *E. M. Forster*, 136. Heard was born in London on 6 Oct. 1889; he died in August 1971. Sadly, like Plunkett, Heard suffered from a long physical decline before his death, caused by more than thirty strokes. Fryer, *Eye of the Camera*, 212; and Davenport-Hines, *Auden*, 118.

[21] Plunkett diaries, 1 Jan. 1927, 21 Feb. 1927, and 8 April 1927.

[22] Plunkett diaries, 31 Aug. 1928; and Horace Plunkett to E. M. House, 3 June 1929 and 20 Aug. 1929, Plunkett collection, Library of Congress.

quite deaf for a number of years, he found it especially annoying to find that Young was also now "almost deaf" as well.) This was the last time Young saw Horace Plunkett.[23]

Near the end of January 1932, Plunkett wrote an eerily prophetic letter to Conrad Young. He had set up procedures to protect Young's interests, and to illustrate how they would work he said, "To put the matter in concrete form, let us suppose that I die on April 1st, 1932." This hypothetical missed the actual event by only five days, for Sir Horace Curzon Plunkett died at Crest House, Surrey, on March 26, 1932.[24]

---

[23] Plunkett diaries, 16 and 27 July 1931 and 26 Jan. 1932.

[24] Horace Plunkett to Conrad Young, 26 Jan. 1932, NSHS collection.

15

The Aftermath

AFTER MANY CHANGES in his testamentary dispositions, Plunkett finally selected two key men to manage his estate: his executor, Gerald Heard, and his man in Omaha, Conrad Young. In a letter to Young, Plunkett emphasized his great confidence in Heard, saying, "A finer character and intelligence than his I have never been fortunate enough to have as a guide, philosopher and friend." He assured Conrad that although Heard was a very busy man, "He has undertaken the task [as executor] out of pure friendship, and would have acted without . . . remuneration."[1]

Heard was also one of the important beneficiaries under the will, under the terms of which he received £1,000 as executor and a further £4,000 to manage the investments in the United States. He also received all of Plunkett's furniture, wearing apparel, and manuscripts and copyrights; in addition, because he was a trustee of the Plunkett Foundation, he could expect compensation from that source, as well. Heard was to have complete authority over the stock in the Nebraska Company, so as to avoid any undue pressure on Young's management of the underlying investments.[2]

One-sixth of the residual estate was to be distributed to the Horace Plunkett Foundation. All of Plunkett's siblings predeceased him, and with one exception, he made no provision for their

[1] Horace Plunkett to Conrad Young, 26 July 1931, NSHS collection.

[2] The £4,000 bequest to Heard was to be paid only from the American assets. A copy of the will and codicil are in the Plunkett collection, NSHS. Gerald Heard, in Digby, *Horace Plunkett,* 297; and West, *Horace Plunkett, Co-operation and Politics,* 200–201. Also, Plunkett diaries, 5 Feb. 1930.

children. He did provide for the children of his sister Mary, and this bequest may have resulted in part from the nasty wrangle he had had with Mary and her daughters over his preferment in income from his father's estate. His share of that estate was disproportionately large, and he felt a moral obligation to provide for his sister Mary's children from the rest of his estate.[3]

Plunkett also provided for the widow and children of his cousin Lord Fingall. Horace said his motive was to save the twelfth century and its traditions for Ireland, but others have suggested he had very different reasons. One writer has bluntly asserted that Plunkett and Daisy Fingall were lovers, and that one or more of her children were his progeny. Considering the deplorable state of Fingall's finances, it is virtually certain that Plunkett financed Daisy's many journeys with him about the British Isles and in Europe. Whatever their domestic situation may have been, it is also clear that Lord Fingall acquiesced in it.[4]

Plunkett grew increasingly concerned that Mary's children might be dissatisfied with their cash legacies and challenge the will, which would upset the structure he had devised for Heard and Young to administer the American assets. To protect that structure he bequeathed the American assets absolutely to Heard, giving him absolute and uncontrolled discretion to sell when he felt it was "expedient," and in the meantime, he had full power to manage those properties. It is clear, however, that Heard understood the will to mean that any cash he received from the American

[3] Plunkett originally allotted another £3,000 to the foundation in his will, but he revoked that amount by the codicil of 16 Feb. 1932, probably because the foundation was not fulfilling its purposes as he conceived them. Of course, he may also have been concerned that his estate might not be large enough to satisfy all of the specific bequests, which subsequently proved to be the case. The Horace Plunkett Foundation was established by a deed from Plunkett dated 17 Jan. 1919. On 27 Jan. 1931, Plunkett mused in his diary that he ought to use the sums planned for the foundation "elsewhere." Plunkett's sister Mary Ponsonby died 5 July 1921. The last point of disagreement between Horace and Mary involved his position on the Irish Question, and thereafter she ceased to write to him. Plunkett diaries, 5 July 1921.

[4] Plunkett diaries, 30 Jan. 1930; and Pethica, *Lady Gregory's Diaries,* 259, note 154. Plunkett was with Daisy at or near the theoretical conception date for each of her children, so it is theoretically possible chronologically for Plunkett to have been the father of all of them. After his brother upbraided him regarding the Daisy Fingall rumors, Plunkett confided in his diary, "D[aisy] & I fully understand each other & F[ingall] understands both." Fingall died in the fall of 1929.

properties should be paid into the trust he administered for the benefit of the legatees under the will.

The gross value of Plunkett's estate was just over £45,000, of which over £30,000 represented assets in the United States, and the net value, after the payment of death duties, was £39,540. Notwithstanding the value declared for the U.S. assets, most of those assets were essentially illiquid and could not be used to satisfy the specific bequests under the will. The remnants of English property were little help in that regard, for the executor was put to considerable expense in selling Plunkett's Crest House and the London apartment, and the cars and other personal effects brought "low sums." The specific bequests, excluding the £4,000 for Heard, amounted to over £18,000, plus whatever amount was needed for the year's salary payments to employees. The required expenditures, including that of Inland Revenue, had to be paid from the cash realized up to that point, and the remainder was only enough to pay 75 percent of the legacies.[5]

When he died, Plunkett still owned a brokerage account in a New York bank, some worthless mining stock, the Diamond Company note, the liquidating certificate from Wyoming Development Company, and the shares of the Nebraska Company. The Nebraska Company owned the shares of the two Wheatland companies (the Wyoming Development Company and the Wheatland Industrial Company) and the Omaha real estate.[6]

Inland Revenue apparently accepted a valuation of $152,000 for the American properties, about half the value Young had given Plunkett a couple of years earlier. Details regarding what made up the total are unavailable, but the real estate controlled by Young in Omaha must have accounted for most of it. The executor declared the shares of the Wheatland companies were worthless for the purpose of the U.S. estate tax and the U. K. death duties. The companies never paid a dividend to the shareholders until 1946, and when Plunkett's Nebraska holding company was liquidated, it

[5] The employees, in addition to those specifically mentioned, included clerk, indoor and outdoor servant, and "any other person" with at least six month's service, so that the number receiving payments might have been half a dozen. Crest House, in Surrey, was Plunkett's home after Kilteragh was burned.

[6] It is uncertain if the brokerage account was owned by the Nebraska company, but clearly Young had some control over it.

was still holding a Wheatland Industrial Company note that had an unpaid principal of $1,991.50, on which interest had accrued at 6 percent for the past thirty-four years.[7]

Conrad Young was eager to have Heard cross the Atlantic to visit Omaha and see the properties there. However, Heard was reluctant to incur the expense for such a journey, noting that the heirs "may be critical, both of each other & of the . . . executor," particularly because it would be some time before they could be paid, "even in part." Heard did not visit Omaha until he moved to the United States in 1937 and stopped in Omaha on his way to California.[8]

Cannon (Plunkett's auditor) told the English solicitors that the Nebraska Company had paid no dividends in the last five years, and Inland Revenue raised the question as to why a property worth over $150,000 yielded so little taxable income in the years Plunkett held it. The estate eventually had to pay taxes on the Nebraska Company dividends, and this outlay further delayed payments to the specific legacies. As four of the legatees badly needed the money and two were in poor health, some hardship resulted. The legatees received a first payment of nearly £560 in the spring of 1943, and this was followed by a further $5,000 in the fall of that year.[9]

In the forties, the pace of liquidation quickened, and Young collected nearly $180,000 in those years. Nevertheless, Plunkett's

---

[7] The Wheatland Industrial Company note was paid in 1945 although without the full interest accrual. Young recovered income of $8,349.93 in 1946, $4,715.37 early in 1947, and a further $7,266.86 later in the year from the Wheatland companies. At the end of 1945, he estimated that a further $5,000 might eventually be paid by these companies. Conrad Young to Edward L. Thorold, 27 Dec. 1945; and H. F. Heard to Conrad Young, 1 Sept. 1945, 28 Feb. 1947, and 20 Nov. 1947, NSHS collection.

[8] Heard, who would neither own a car nor drive one, came to the United States with the Huxleys and traveled across the country in their car with them. Heard apparently visited both the Omaha properties and the Wheatland Project. Bedford, *Aldous Huxley*, 335; and Grover Smith, *Letters of Aldous Huxley*, 422. Also, H. F. Heard to Conrad Young, 23 Aug. 1932, NSHS collection.

[9] In the spring of 1940, the Nebraska Company finally had enough funds to pay $9,000 on Heard's specific legacy of £4,000, which he had not yet claimed from the estate. The following year, the company accumulated a further $5,000, which was also paid to Heard. The British solicitors insisted that Heard receive interest on the delayed legacy, although he was willing to waive it in favor of the other legatees; he received £420 as a result. During the time when the legacy was unpaid, the British pound devalued to $4.02, which reduced his legacy, denominated in pounds. H. F. Heard to Conrad Young, 30 April 1940, 12 Aug. 1941, 2 Oct. 1941, 28 Oct. 1943, 8 May 1943, and 18 Sept. 1943, NSHS collection.

estate was still not settled, for the British income tax problem reverberated in the United States, where further taxes and penalties were also assessed. The remaining cash of nearly $10,000 was only sent over to England in 1949, after the statute of limitations expired on all the outstanding tax disputes. The English solicitor for the estate told Heard that the probate was the most complex he had ever handled.[10]

When the assets managed by Young had all been liquidated, the estate realized at least $200,000, not the $152,000 reported to Inland Revenue at the time of Plunkett's death. Nevertheless, payments by the Wyoming Development Company and the Wheatland Industrial Company of just over $30,000 were a steep discount for the $193,000 liquidation certificate Plunkett received in 1914. Liquidating Plunkett's estate took so much time that Young, who was a successful Omaha realtor by this time, also retired in 1945 at the age of 71. He had come to work for Plunkett in 1904 and had spent over forty years dealing with the Irishman's assets.

---

[10] In 1944, securities sales funded a $5,000 payment of legacies, and real estate sales yielded a further $6,464.87. The following year, a total of $138,000 came from real estate. In 1946–47, the Wheatland companies paid a total of $20,265.30, and the sale of the liquidation certificate yielded another $10,000. H. F. Heard to Wyoming Development Co., January 1948; Conrad Young to Edward L. Thorold, 2 March 1949; and H. F. Heard to Conrad Young, 25 March 1949 and 9 Dec. 1949, NSHS collection.

Epilogue

ALTHOUGH THE FINAL $200,000 Young was able to realize from Plunkett's estate was a substantial amount of money, it was a poor return on Plunkett's investment in time and effort, and one must look elsewhere for the American fruits of his life's work.

Certainly, his inquisitive mind never tired of seeking out and learning new ideas. Sometimes this quest for knowledge could be carried to a fault. Indeed, late in life he remembered that a cowboy in Wyoming once said of him, "He can't keep his mind long enough on one thing to boil an egg." Plunkett also would have wanted to be credited with the ability to see both sides of an issue, even though his maddening insistence on being even-handed—according to his definition of that term—too often made enemies of the people who otherwise would have been his friends.[1]

His management skills were considerable, and he used the sort of forward planning that we now regard as essential to business success. In particular, his efforts contributed significantly to the final success of the Wheatland Project, and the Diamond Ranch also owed much to him. Yet, he must be faulted for his frequent failure both to find compatible partners and to hire the best people for the job; the history of his failed ventures is filled with examples of flawed business relationships.

Others have remarked that Plunkett's American adventures molded his future life in significant ways, and that was certainly

[1] Plunkett diaries, 16 Dec. 1926.

true, although there are not many specific examples that can be adduced. In the ten years of his ranching career, he spent over half of his time in the American West, and the passing of that era did not mark the end of his love affair with the United States. In the next thirty-five years, he crossed and re-crossed the ocean more than sixty times, spending an average of almost two months in America each time. In all, he a spent a total of nearly ten years of his adult life in the United States.[2]

The American cooperative movement owes much to Horace Plunkett, as he was instrumental in raising the first $35,000 to establish the American Agricultural Organizing Society. Certainly Charles McCarthy, who adapted Plunkett's concepts of the cooperative movement to conditions in the United States, always acknowledged his debt to the Irishman. Although his achievement in this field—as with all the things he wanted to do—fell short of his desire, the fruits of his labors for American cooperation were impressive.[3]

Remarkably, he had access to many American leaders, both in business and in government, and at the high point of his work in America he was an intimate advisor to presidents. The range of interests that brought him to the United States was very broad, encompassing ranching, agricultural cooperatives, medicine, promotion of Irish manufactures, and, of course, foreign policy. Rarely do we find one person with knowledge and competence in so many fields.

---

[2] Plunkett made more than 80 trips back and forth across the Atlantic.

[3] Fitzpatrick, *McCarthy of Wisconsin*, 186.

Millais Photographs

THE TWENTY PHOTOGRAPHS reproduced here were taken by Geoffroy William Millais (younger son of the famous artist Sir John Everett Millais), while he was ranching in Wyoming's Powder River Basin. Millais was ranching with Thomas Willing Peters and Walter C. Alston, and a number of the photos are of their Bar C Ranch. Millais was a near neighbor of Horace Plunkett in the Powder River Basin, and some of the (unidentified) cowboys in the photographs are likely Plunkett employees. A number of these photos are not otherwise identified, but all are authentic records of ranchers and their cowhands on the Wyoming ranges when Plunkett was there in the 1880s (most were probably taken in October 1887). Courtesy of the American Heritage Center, University of Wyoming; Margaret Brock Hanson papers; and also Mrs. E. G. Millais and the family of Edward Gray St. Helier Millais (son of Sir Geoffroy William Millais, who died in 1941).

Peters and Alston's Bar C Ranch house
Although far less ostentatious than Frewen's "castle," the Bar C Ranch house featured the luxury of a husband and wife who cooked and provided creature comforts for the men.

Another view of the Bar C Ranch house.

Bar C Ranch foreman Henry W. Devoe with his horse.

Barn at Bar C Ranch
(*standing at left*) foreman Henry W. Devoe; (*center*) Devoe's son Earl; (*standing with baby*) Devoe's wife, May A. Devoe, holding daughter Lydia.

Ranch hands at the Bar C
Tentative partial identification: (*man kneeling at left*) Jimmy Rennick; (*man kneeling second from left*) Thomas F. Carr.

Roundup in the Hole in the Wall, Wyo.
The Hole in the Wall, famous outlaw hideout, is a valley between the Big Horn Mountains on the west and the fifty-mile Red Wall, which is broken in the north by one deep canyon near Peters and Alston's Bar C Ranch.

Cowhands at the chuck wagon,
Red Wall in the background
Partial identification: (*third from left*) Charles Morgareidge;
(*far right*) Thomas F. Carr.

Steamboat Rock,
landmark in the Hole in the Wall, at the north end
of the Red Wall, near Peters and Alston's Bar C Ranch.

Another view of Steamboat Rock.

Cunningham Flat,
south of Peters and Alston's Bar C Ranch, Hole in the Wall area.

Roundup scene,
perhaps near Rough Lock Hill,
at the south end of the Hole in the Wall area.

Large group of ranch hands, probably roundup scene.

Ranch hands taking a break near chuck wagon,
with Red Wall in background.

Ranch hands with chuck wagon and bedrolls.

Ranch hands by chuck wagon.

Frewen's "castle," headquarters of Moreton Frewen's Powder River Cattle Co. on the Powder River, ca. 1884–87
The house featured a forty-foot room with a mezzanine, where a piano provided music for guests below, and a telephone gave instant access to information from the store at Powder River crossing, twenty-four miles away.

Along Powder River,
perhaps near Powder River Crossing (Old Ft. Reno).

Antlers at hunting campsite.

Unidentified log frame ranch house.

Unidentified log house with sapling.

Bibliography

Manuscript Collections

Bosler collection. American Heritage Center. University of Wyoming.
Brayer collection. American Heritage Center. University of Wyoming.
Carey collection. American Heritage Center. University of Wyoming.
Frewen, Moreton, Collection. American Heritage Center. University of Wyoming.
Hesse, Fred G. S., Collection. American Heritage Center. University of Wyoming.
Ione Land & Cattle Company. Ione collection. Western Range Cattle Industry Study. Colorado Historical Society
Plunkett, Sir Horace Curzon. Diaries. Library of Congress. [Horace Plunkett's diaries, dating from 1885 until just before his death in 1932, are the most important source for this work. Microfilm copies can be found at the Library of Congress.]
Plunkett, Sir Horace Curzon. Plunkett collection. Nebraska State Historical Society. [This collection contains information on Plunkett's Omaha investments.]
Western Range Cattle Industry Study. Library of Congress. [The WRCIS microfilm is a very useful source for the British cattle companies.]
Wyoming Development Company Minutes. Carey Collection. American Heritage Center. University of Wyoming.

Books and Journals

Anderson, R. A. *With Horace Plunkett in Ireland.* London, 1935.
Bedford, Sybille. *Aldous Huxley: A Biography.* New York, 1973.
Benson, Arthur C. *Memories and Friends.* London, 1924.
Black, Edwin. *War against the Weak: Eugenics and America's Campaign to Create a Master Race.* New York, 2003.

Bolotin, Norman, and Christine Laing. *The World's Columbian Exposition: The Chicago World's Fair of 1893*. Washington, D.C., 1992.

Brecher, Ruth, and Edward Brecher. *The Rays: A History of Radiology in the United States and Canada*. Baltimore, 1969.

Burns, Robert, Andrew Gillespie, and Willing Richardson. *Wyoming Pioneer Ranches*. Laramie, Wyo., 1955.

Burroughs, John Rolfe. *Guardian of the Grasslands: The First Hundred Years of the Wyoming Stock Growers Association*. Cheyenne, Wyo., 1971.

Carver, Thomas Nixon. *Principles of Rural Economics*. New York, 1911.

Chernow, Ron. *Titan: The Life of John D. Rockefeller, Sr.* New York, 1998.

Clay, John. *My Life on the Range*. New York, 1961.

Cohen, Naomi W. *A Dual Heritage: The Public Career of Oscar S. Straus*. Philadelphia, 1969.

Davenport-Hines, Richard. *Auden*. New York, 1995.

Digby, Margaret. *Horace Plunkett: An Anglo-American Irishman*. Oxford, 1949.

Dustin, Dorothy Devereux. *Omaha & Douglas County: A Panoramic History*. Woodland Hills, Calif., 1980.

Ferrill, Robert H. *American Diplomacy: A History*. New York, 1969.

Fitzpatrick, Edward A. *McCarthy of Wisconsin*. New York, 1944.

Frison, Paul. *First White Woman in the Big Horn Basin*. Basin, Wyo., 1985.

Fryer, Jonathan. *Eye of the Camera: A Life of Christopher Isherwood*. London, 1993.

Furbank, Philip Nicholas. *E. M. Forster: A Life*. New York, 1977.

Gordon, Archibald Victor Dudley, Marquess of Aberdeen. *A Wild Flight of Gordons*. London, 1985.

Gosnell, Harold F. *Boss Platt and His New York Machine: A Study of the Political Leadership of Thomas C. Platt, Theodore Roosevelt and Others*. New York, 1924.

Gressley, Gene M. *Bankers and Cattlemen*. New York, 1966.

Grey, Edward, Viscount of Fallodon. *Twenty-Five Years, 1892–1916*. 2 vols. New York, 1925.

Hardy, Deborah. *Wyoming University: The First 100 Years, 1886–1986*. Laramie, Wyo., 1986.

Harnack, Curtis. *Gentlemen on the Prairie*. Ames, Iowa, 1985.

Hart, John Mason. *Empire and Revolution: The Americans in Mexico since the Civil War*. Los Angeles, 2002.

Hendrick, Burton J. *The Life and Letters of Walter H. Page*. 3 vols. New York, 1923, 1926.

Hinkson, Pamela, ed. *Seventy Years Young: Memories of Elizabeth, Countess of Fingall*. Dublin, 1991.

Hodgson, Godfrey. *Woodrow Wilson's Right Hand: The Life of Colonel Edward M. House.* New Haven, Conn., 2006.

Ishbel, Countess of Aberdeen. "Ireland at the World's Fair." *The North American Review* 157 (July 1893): 18–24.

———. *We Twa.* 2 vols. London, 1925.

Kaplan, Justin. *When the Astors Owned New York: Blue Bloods and Grand Hotels in a Gilded Age.* New York, 2006.

Kavanagh, Art. *The Landed Gentry & Aristocracy: Meath.* Dublin, 2005.

Keen, Elizabeth. "Wyoming's Frontier Newspapers." *Annals of Wyoming* 34, no. 1 (April 1962): 65–66.

Knock, Thomas J. *To End All Wars: Woodrow Wilson and the Quest for a New World Order.* Princeton, N.J., 1992.

Kuhn, William Marshall. *Henry and Mary Ponsonby: Life at the Court of Queen Victoria.* London, 2002.

Larson, Olaf F., and Thomas B. Jones. "The Unpublished Data from Roosevelt's Commission on Country Life." *Agricultural History* 50, no. 4 (1976): 583–99.

Larson, Taft Alfred. "The Winter of 1886–87 in Wyoming." *Annals of Wyoming* 14, no. 1 (Jan. 1942): 5–17.

Leech, Harper, and John Charles Carroll. *Armour and His Times.* Freeport, N.Y., 1938.

Leslie, Anita. *The Marlborough House Set.* New York, 1973.

———. *Mr. Frewen of England: A Victorian Adventurer.* London, 1966.

Link, Arthur S., ed. *Papers of Woodrow Wilson*, vol. 33. Princeton, N.J., 1980.

———. *Papers of Woodrow Wilson*, vol. 35. Princeton, N.J., 1981.

———. *Papers of Woodrow Wilson*, vol. 40. Princeton, N.J., 1982.

Longford, Elizabeth. *Queen Victoria: Born to Succeed.* New York, 1964.

Lott, Howard B. "The Old Occidental." *Annals of Wyoming* 27, no. 1 (April 1955): 27.

Lysaght, Edward E. *Sir Horace Plunkett and His Place in the Irish Nation.* Dublin, 1916.

Magnus, Philip. *King Edward the Seventh.* New York, 1964.

Marilley, Suzanne M. *Woman Suffrage and the Origins of Liberal Feminism in the United States, 1820–1920.* Cambridge, Mass., 1996.

Marshall, Susan E. *Splintered Sisterhood: Gender and Class in the Campaign against Woman Suffrage.* Madison, Wis., 1997.

Martin, Albro. *James J. Hill and the Opening of the Northwest.* St. Paul, Minn., 1991.

Martinex-Brawley, Emilia E. *Seven Decades of Rural Social Work: From Country Life Commission to Rural Caucus.* New York, 1981.

McCarthy, Charles. *The Wisconsin Idea.* New York, 1912.

McGurrin, James. *Bourke Cockran: A Free Lance in American Politics.* New York, 1948.

Morris, Edmund. *The Rise of Theodore Roosevelt.* New York, 1979.

Nye, Frank Wilson. *Bill Nye: His Own Life Story.* New York, 1926.

Olson, James C. *History of Nebraska.* Lincoln, Neb., 1966.

Omaha Stockyards: A Century of Marketing, 1884–1984. Omaha, Neb., 1984.

Osborne, Meta. "John Alexander Osborne." *Annals of Wyoming* 14, no. 4 (Oct. 1942): 315–21.

Pethica, James, ed. *Lady Gregory's Diaries 1892–1902.* New York, 1996.

Pinchot, Gifford. *Breaking New Ground.* New York, 1947.

Plunkett, Sir Horace. "Better Farming, Better Business, Better Living." *The Outlook* 94, no. 8 (19 Feb. 1910): 397–401.

———. "Better Farming, Better Business, Better Living: Two Practical Suggestions" *The Outlook* 94, no. 9 (26 Feb. 1910): 497–502.

———."Conservation and Rural Life: An Irish View of Two Roosevelt Policies." *The Outlook* 94, no. 5 (29 Jan. 1910): 260–64.

———. "The Human Factor in Rural Life." *The Outlook* 94, no. 7 (12 Feb. 1910): 354–59.

———. *Ireland in the New Century.* London, 1904.

———. "The Irish Question in a New Light." *North American Review* 166 (Jan. 1898): 107–120.

———. *The Rural Life Problem of the United States: Notes of an Irish Observer.* New York, 1910.

———. *Some Tendencies of Modern Medicine—From a Lay Point of View.* Battle Creek, Mich., 1913.

———. *The Working of Woman Suffrage in Wyoming.* Cheyenne, Wyo., 1890.

Report of the Commission on Country Life, With an Introduction by Theodore Roosevelt. New York, 1975. First published 1911 by Sturgis & Walton.

Riley, Noël. *Tile Art: A History of Decorative Tiles.* London, 1987.

Robbins, Roy M. *Our Landed Heritage: The Public Domain 1776–1936.* Lincoln, Neb., 1962.

Robinson, Lennox, ed. *Lady Gregory's Journals.* New York, 1947.

Schwarz, Richard W. *John Harvey Kellogg.* Nashville, Tenn., 1970.

Seymour, Charles, ed. *The Intimate Papers of Colonel House.* 4 vols. New York, 1928.

Smith, Grover, ed. *Letters of Aldous Huxley.* New York, 1969.

Smith, Helena Huntington. *The War on the Powder River.* Lincoln, Neb., 1966.

Smith, Page. *America Enters the World: A People's History of the Progressive Era and World War I.* New York, 1985.

Smith, Richard Norton. *An Uncommon Man: The Triumph of Herbert Hoover.* Worland, Wyo., 1984.

Sokolow, Jayme A. *Eros and Modernization: Sylvester Graham, Health Reform, and the Origins of Victorian Sexuality in America.* Cranbury, N.J., 1983.

Stapylton, H. E. C. *Second Series of Eton School Lists, Comprising the Years between 1853 and 1892, with Notes and Index.* London, 1900.

Stern, Alexandra Minna. *Eugenic Nation: Faults and Frontiers of Better Breeding in Modern America.* Los Angeles, 2005.

Tabulation of Adjudicated Water Rights of the State of Wyoming: Water Division Number One. Cheyenne, Wyo., 1965.

Tabulation of Adjudicated Water Rights of the State of Wyoming: Water Division Number Two. Cheyenne, Wyo., 1963.

Townsend, Everett. "Development of the Tile Industry in the United States." *The Bulletin of the American Ceramic Society* 22, no. 5 (15 May 1943), 130.

Tuchman, Barbara W. *The Guns of August.* New York, 1962.

U.S. Senate. Committee on Agriculture and Forestry. *Hearing Relative to Cooperative Agricultural Associations.* 67th Cong., 4th sess., 10 Jan. 1923.

Van Wingerden, Sophia A. *The Women's Suffrage Movement in Britain, 1866–1928.* London, 1999.

Vickers, Hugo. *Royal Orders: The Honours and the Honoured.* London, 1994.

Walker, Tacetta B. *Stories of Early Days in Wyoming: Big Horn Basin.* Casper, Wyo., 1936.

A Week at the Fair: Illustrating the Exhibits and Wonders of the World's Columbian Exposition. Chicago, 1893.

Welsh, Donald H. "Pierre Wibaux: Cattle King." *North Dakota History* 20, no. 1 (Jan. 1953): 5–23.

West, Trevor. *Horace Plunkett, Co-operation and Politics: An Irish Biography.* Washington, D.C., 1986.

Willcox, James K. Hamilton. *Wyoming: The True Cause and Splendid Fruits of Woman Suffrage There, From Official Records and Personal Knowledge; Correcting the Errors of Horace Plunkett and Professor Bryce, and Supplying Omissions in the History of Mrs. Stanton, Mrs. Gage and Miss Anthony, and in the History of Wyoming by Hubert Howe Bancroft; With Other Information About that State.* New York, 1890.

Woods, Lawrence M. *British Gentlemen in the Wild West: The Era of the Intensely English Cowboy.* New York, 1990.

———. *Moreton Frewen's Western Adventures.* Boulder, Colo., 1986.
———. *Sometimes the Books Froze: Wyoming's Economy and Its Banks.* Boulder, Colo., 1985.
———. *Wyoming Biographies.* Worland, Wyo., 1991.
Wortham, H. E. *Victorian Eton and Cambridge: Being the Life and Times of Oscar Browning.* London, 1956.
Yost, Nellie Snyder. *The Call of the Range: The Story of the Nebraska Stock Growers Association.* Denver, Colo., 1966.

Index

HCP = Horace Curzon Plunkett
References to photographs are in italic type.